STAY ALIVE

WHILE DRIVING!

THE SECRETS
OF A
RACE CAR DRIVER

What you *need* to know to protect you and your loved ones from the craziness that is the public streets and highways.

LOREN ELMER

WTSI PUBLICATIONS
SANTA MONICA, CALIFORNIA
2015

SPECIAL ACKNOWLEDGEMENTS

I want to give a big THANK YOU for their help and support to BARBARA ANDRISANI, GLORIA FOX, ELIZABETH DOBLE, my brother LON ELMER, RON FIENI, DAVE CARMAN, and JACK BAZELLA who accidently started this whole thing.

Acknowledgement is made to the following for their kind permission to reprint copyrighted material: Excerpted from *The Unfair Advantage* by Mark Donohue, © Bentley Publishers, Cambridge, MA, www.bentleypublishers.com, ISBN 0-8376-0069-3.

This WTSI Publications edition is the first publication of
STAY ALIVE WHILE DRIVING
THE SECRETS OF A RACE CAR DRIVER.

WTSI Publications edition: 2015

ISBN 9780578166100
Copyright Office Registration Number TXu 1-973-246

PRINTED IN THE UNITED STATES OF AMERICA
FIRST EDITION

CONTENTS

INTRO

You're driving along a country road. You round a corner and hit a patch of gravel. Your car goes into a skid. What's the last thing you see before you hit the tree?

...The tree.

The reason you hit the tree?

Because you were looking at it.

- - - - - - - - - -

In the movie LE MANS, Steve McQueen's character had just been released from the trackside hospital after surviving a crash during the twenty-four hour race. The widow of a driver he had raced against began questioning him about the obvious inherent risks and without saying so was trying to understand what drove her husband ultimately to his death. She finally asked, "What is so important about driving faster than everyone else?" After some thought Steve McQueen answered, "When you're racing, it's life. Anything that happens before or after...is just waiting."

Racecar driving is incredibly intoxicating, the speed, the risk, and the resulting adrenalin rush. It is romanticized, glorified; we view it as something we all would like to do. Well, some of us anyway. But you have to admit, racecar driving ranks right up there with the all-time top thrill-seeking sports. Hemmingway once said there were only three sports: mountain climbing, bullfighting and auto racing. Everything else he considered just a game. His distinction being that with a *real* sport...any miss-step meant death.

So what's all this stuff about the Secrets of a Race Car Driver? What does race driving have to do with everyday driving, driving to work or driving to the store?

Well, when an irresponsible driver is texting and blows through a red light, when a kid runs in front of your car, when you skid in a corner and you're headed for that tree, in those split seconds, racecar driving has everything to do with you and your everyday driving. What is the difference between race driving and you suddenly trying to avoid a collision?

NOTHING. There is NO DIFFERENCE.

A race driver is asking the car to give him everything it's got to get him around the track as quickly as possible. When you swerve to avoid a collision you are asking *your* car to give you everything it's got to get you out of that situation. When you push a car to its limits asking for everything it's got, whether racing or avoiding a collision, there is no difference between you and a racecar driver. In those split seconds the driving techniques a race driver needs to know to succeed are the same driving techniques you need to know to survive. And they are the SECRETS of a RACECAR DRIVER.

I taught race schools and I taught emergency driving schools where you learn what to do when a drunk driver comes at you head-on, when a kid runs in front of your car, or when your car skids. The techniques I taught in the emergency driving schools were the exact same techniques I taught in the racing schools. Because racing is nothing more than continuous emergency driving.

We all know how to drive. You start the car, put it in gear, take your foot off the brake, accelerate, turn the wheel for corners, and stop when you get to where you're going. A simple enough process, right? And yet every day doing that same simple process, people are getting killed.

The roads are more crowded; cars are going faster; and apparently the only requirement for a driver's license is a pulse. Every day we see people doing things behind the wheel that defy logic and just plain common sense, endangering everyone around them as well as themselves: using their knee for steering since their sandwich requires two hands, dialing the cell phone with an occasional glance in the rear view mirror to see if they hit something, driving while intoxicated and, of course, texting. If they can text really fast doesn't mean they'll be looking at the road more often. Just the opposite: they'll get in *more* texting while NOT looking at the road, oblivious of red lights, pedestrians, and that thump when their car bounced over something. This book is about the very specific driving techniques you need to know to avoid those drivers. And they are the same techniques race drivers use to go fast and stay alive.

Now the techniques of racecar driving will at first seem simple. But doing the techniques to where they are instinctive when that tree or that irresponsible driver is getting frighteningly close is quite another matter. It's the old "easier said than done." They take understanding and practice, the over-and-over-again practice that establishes motor memory. But you don't have to be on a racetrack to do this, "seat time" as we say in racing. This practice can be done everyday as you go about your daily driving. Yes, you're daily driving.

HOWEVER!

I want to be perfectly clear about this: everything I'm teaching you in this book can and is to be done AT or BELOW the NORMAL SPEED LIMIT. I don't want you joining that select group of drivers who are trying for the Darwin Award - an award given each year to those individuals who only through luck and God's good grace managed to survive their own stupidity. The award is given posthumously to those who ran out of luck and God was looking elsewhere at the time.

If you think you can fly, you don't jump off a ten-story building. You jump off nothing higher than a curb. Just because the techniques are the Secrets of a Race Car Driver, doesn't mean you need to be speeding to do them. Just the opposite, you'll learn the techniques faster if you are at a slower speed. When I'm coaching my clients on a racetrack, I slow them down so they'll work on the technique. Speed at that point is just a distraction that keeps them from learning. Once the clients get the technique down, the speed will come and far exceed anything they were doing before. In the case of your driving on the public streets, when you DON'T SPEED these techniques are more quickly learned and once mastered can be the difference between hitting or not hitting the tree, and more often than not the difference between life and death.

A study was done years ago to see if racecar drivers with their advanced driving skills had fewer accidents on the public streets and highways. As it turns out, they had more. One can understand, not necessarily forgive, but understand racecar drivers tend to drive more aggressively. Couple that with a more cavalier attitude about trading fenders and you see where racecar drivers would have more accidents. But the study also found something else. Racecar drivers had... *fewer fatalities*.

By learning the techniques in this book and by practicing these techniques in your everyday driving, you will be better prepared to survive the rigors of crowded roads, irresponsible drivers and even the laws of physics.

You will learn that THE SECRETS OF A RACE CAR DRIVER are what you need to know to STAY ALIVE WHILE DRIVING.

MY FIRST

It was during a coffee break at work that Jack pulled his nose out of the newspaper to say, "Hey, look here there's a race car for sale."

He immediately had my attention.

"It's a Formula Car. Some guy in town has one for sale. What do you say you and I go together on it? We can take turns racing it."

The air was going out of the tires fast.

"Right Jack, you with two daughters in private school? You can barely afford gas let alone the maintenance on a racecar. Besides, your wife would kill you if the racecar didn't."

But I had to admit I was curious.

What the heck is a Formula Car?

The guy swung the garage door open and there she was.

It was love at first sight.

She was low slung and I do mean low slung; the driver seat was only inches off the ground. She was completely stripped down, no extra anything, no fenders, no headlights, certainly no heater or radio. It was built for one thing and only one thing…racing. The driver seat was in a layback position almost like a recliner. It was so low you looked between the front tires, not over them. There was a roll over hoop right behind the seat and behind that was the engine mid-engine is what they called it just in front of the rear axle. She was black and orange with numbers painted on the sides.

I was trying hard not to salivate.

"Well, I'm interested," I told the owner. "But I don't buy any car unless I drive it first."

"No problem," he said.

We pushed it onto a trailer and towed it down to an empty warehouse parking lot. We rolled the car off the trailer and with a wave of his hand, the owner said, "Have at it."

It was Christmas and I was five years old again. I didn't even bother with the six-point harness when I slid into the seat. Made up of lap belts, shoulder belts and a crotch belt, the harness was designed to keep you in the car in case you ended up on your lid.

I just pushed them aside, hit the start button, waited for the oil pressure and some temp and then floored that sucker. In no time I was screaming around the parking lot higher than the proverbial kite. No helmet, no goggles, I could have been blinded by all the crap that was flying off the tires. Sideways through the turns and my hair flattened by the speed, I was in heaven.

Concerned, the owner turned to Jack and asked, "Has he ever driven one of these before?"

Watching me as I was going crazy out there in the parking lot, Jack casually answered, "Nope."

In a cloud of dust I slid to a stop right up to them, hopped out and proudly announced, "I'll take it."

And thus began a life long obsession.

CAR SAFETY

It wasn't all that long ago that talking about car safety was like telling a kid at the dinner table the food he had been staring at for the longest time was actually "good" for him.

What are seat belts? The car manufacturer Tucker was going to introduce seat belts in his cars back in 1947 but was advised against it. His design team thought it sent the wrong message.#1 He went out of business. Not, of course, because of the seat belts but you get the idea: seat belts were considered more a detriment to image than an issue. But as more cars hit the road and started hitting each other, seat belts and car safety did become an issue. That got carried home when highway police organizations began not only keeping records of the number of fatalities on the highway but also began announcing them. Holiday weekends were always followed by the Monday night news announcing the national death toll on the highways.

It gradually began to sink in to car manufacturers that car safety was actually a selling point. Not that the manufacturers weren't aware of car safety all along. They were developing safety along with creature comforts and good looks from the beginning. It just wasn't a selling point and so didn't get as much attention. As far back as the 1920's and 30's the manufacturers were doing things like switching cars from mechanical brakes to hydraulic brakes and greatly improving the chances of avoiding a collision.

But looking at car posters from the fifties and sixties, creature comfort and external beauty were the selling points, not safety. Those big old beauties seemed to just float down the road. There was so much power in the power steering when you turned the wheel you had no feel for the road whatsoever; you just floated around the corner.

By the late sixties and early seventies, speed and excitement became the new advertising mantra ushered in by those beautiful beasts, the Muscle Cars. More horse *power* than horse sense, Muscle Cars were an acceleration orgy. Mash the gas pedal to the floor and you were instantly shoved back deep into your seat. Yeehaw, what a sensation! However there was a problem. If you had to turn or stop… Many a telephone pole was cut in half by airborne Muscle Cars that had no chance of making the corner.

It was racing that began to change all that. The Trans-Am Championships pitted those same Muscle Cars in competition. The better suspensions and bigger brakes developed for the race series eventually found their way to the showroom floor. As it's often said, "If you're going to have lots of go, you better have lots of whoa." An entire book could be written just on the advancements in car innovation that came over from racing. Anything on the leading edge is always going to have a trail of marketable products behind it.

Today, car safety is of primary importance. This is particularly true for new families and parents of teens. I've included a whole chapter devoted to just teen driving, and showing parents how to teach their teens to drive and survive without it turning into a battle. But for everyone, safety has become a major consideration when purchasing a vehicle. People are now asking the crash ratings of a particular vehicle. It's not just the CARFAX® you want to know about. The National Highway Traffic Safety Administration (NHTSA) and the Insurance Institute of Highway Safety (IIHS) are the main organizations that rate car safety, five stars being the top rating. You want to know the rating for front, side and rear impact as well as whether the roof will support the weight of the vehicle in the case of roll over.

This more prevalent interest by the public in car safety has the Federal Government placing mandates on car construction and the manufacturers have responded. All vehicles are designed now with stronger pillars in case of roll over. Some convertibles are designed with extra strong windshield supports, the A pillar, that act as roll over bars. I know a lot of people complain about the A Pillar blocking their vision but there is a reason those things are fatter now. If the car rolls over, it's designed to keep you from getting up close and personal with your floor mat.

I'm reminded of a woman whose Volvo was crushed by a truck. Emergency crews used the Jaws of Life to cut the car from around her. Police were shocked to find she was still alive and suffered only a broken leg. I met her when she was wheeled into a Volvo dealership her leg still in a cast. She came to buy the same model Volvo, the vehicle that saved her life. The strides that *all* manufacturers have made in car safety have been nothing short of impressive. And no, I don't work for Volvo.

There are two safety systems I want to point out, not necessarily from racing, worth mentioning before we get into the Secrets. I'll be referring to these systems throughout. You've heard these terms before because these safety systems have been around for a while; but I'm surprised by how few people actually understand these systems and can recognize them when they activate.

A.B.S.

The Anti-lock Brake System, or as I like to say - Allows you to Brake and Steer - was first developed back in the late 1920's to help stop airplanes on wet runways. It was designed to be an intermittent brake application, the equivalent of pumping the brakes in an attempt to stop under marginal conditions.#2

By the 1970's more resources were being put into understanding car accidents and car safety. When engineers began to realize most people's knee-jerk (read: panic!) reaction to an impending collision was to slam on the brakes and turn the wheel, A.B.S. was the perfect solution.

Here's the problem. When you slam on the brakes you lock up the wheels and the tires loose traction as you skid straight into what you were trying to avoid. If you then add steering to the mix it just makes it worse. Now, you're asking the front tires to do two things at once - stop *and* steer - when the tires couldn't find the traction to just stop. With the front tires locked a car is not going to steer in a different direction. The front tires have to be rolling freely to allow the tires to find the necessary traction to redirect the car. With the brakes locked, the wheel turned and your tires screeching for a lack of traction, you're just watching your life pass before you eyes, usually followed by an expletive or an invocation to a higher being.

A.B.S. however prevents the brakes from locking and is designed to separate people's panic-driven attempt at stopping and steering at the same time. It's the equivalent of pumping the brakes but very, very quickly. It grabs the brake to slow you down just short of skidding. It then releases the brake allowing the front tires to roll and thereby steer. It's asking the front tires to just do *one* thing at a time. It again grabs the brake slowing the vehicle, and again releases it allowing the tires to roll freely so you can steer. It does this eighteen times per second, faster than you could ever pump. When you slam on the brakes and turn the steering wheel to avoid a collision, you get brake-steer-brake-steer-brake until you come to a stop hopefully without hitting anything.

In the early versions of A.B.S., the pulsing in the brake caused a violent pulsing in the brake pedal. People experiencing it for the first time were surprised by it and often took their foot off the brake pedal…and crashed. Since then, engineers have gotten the brake pedal quieted down, and with the advent of electronically applied braking systems you won't feel anything in the pedal at all. However you will feel it in the suspension and subsequently the seat of your pants accompanied by a lot of GRUNTING noise. For those of you in Northern climes experiencing A.B.S. on ice or snow, the noise and the jerking won't be as violent. But on a dry pavement the noise and the stuttering sensation in the seat of your pants can freak out first time users, especially teens. Teens will take their foot off the brake pedal thinking they did something *wrong*. The same reaction they have when you yell their full name. So, find a BIG EMPTY parking lot and test the A.B.S. brakes so you won't be surprised when you do have to use them. Again you don't have to be speeding; thirty MPH tops with lots of room and you'll be good. The key words here are BIG and EMPTY parking lot. Avoid the ones with lots of light poles or median strips with trees or those little concrete parking space designators. Do NOT do this on the street. The guy behind you will *not* be expecting your sudden arbitrary stop. He's texting and he'll slam into you. It's for panic stopping; so don't wait for a panic situation to find out what it feels like. Try it out, and try the braking and steering together so you know how that feels as well.

Also, a number of people are *not* aware that A.B.S. is a POSITION on the brake pedal. You have to push the brake pedal all the way down to the lower one third of brake pedal travel. In other words, shove the brake pedal to the floorboard. That will activate the system causing the brake pedal to pulse on and off allowing you to stop and steer at the same time to avoid a collision. Often people think if you just hit the pedal hard or fast, you'll activate the A.B.S.. But that's not enough. That's why some manufacturers have added a system called Brake Assist. It recognizes when the person is panic stopping. If the driver doesn't push the brake pedal far enough, Brake Assist will *pull* the brake pedal down to A.B.S.

When first introduced, people came back to the dealerships complaining of brake problems – the brake pedal had gone suddenly to the floorboard. When asked if the car had stopped and the answer was yes they had stopped quite suddenly…they had triggered the Brake Assist.

The question of course is: how does Brake Assist know you are panic stopping as opposed to just normal stopping? It knows by sensing how quickly your foot goes from the accelerator pedal to the brake pedal and how quickly you push the brake pedal. In a normal stop you can count, "one-thousand-one," going from the gas to the brake. In a panic stop, you'll be lucky to just say, "one," before hitting the brake pedal.

Again, in a panic stop push the pedal all the way down, and with the ball of your foot, not with your toes as when stretching to reach the pedals. You do have to push the pedal hard to activate A.B.S.

Sit close enough so you can push the brake pedal firmly to the floorboard. If you don't you'll be just braking, and if you add steering you'll be asking the tires to do *two* things at once – stop *and* steer - which in a panic stop could result in a skid. A.B.S. is designed to ask the tires to do just *one* thing at a time – stop *then* steer *then* stop *then* steer *then* stop, greatly increasing your chance of avoiding harm.

STABILITY CONTROL

The other system I want to talk about is Stability Control, one of the best auto safety innovations in the last couple of decades. It comes under a number of different manufacturer names: Dynamic Stability Control System or Program, Electronic Stability Control, or Vehicle Dynamic Stability, but they all refer to the same thing. They are a system designed to help you regain control of the vehicle when it SKIDS OFF your INTENEDED PATH.

There are only two types of skids: OVER STEER, or "Loose" as they say in racing, where the REAR TIRES lose traction and swing out to try to pass the front tires. The other type of skid is UNDER STEER, or "Push," where the FRONT TIRES lose traction and don't allow you to turn the vehicle.

To intervene and assist you in regaining control and continuing on your Intended Path, the Stability Control uses a yaw sensor to recognize the skid. In the case of Over Steer where the rear tires lose traction, the yaw sensor recognizes the vehicle is skidding sideways into a rotation instead of staying on the Intended Path as the rear tires skid to one side or the other. This triggers the Stability Control to cut the engine power. We don't want to be asking those tires to do two things at once: find traction *and* accelerate. With just one thing at a time, the tires are more capable to respond. The Stability Control then activates the A.B.S. - the Anti-Lock Brake System - to grab a *single* brake. In this case it will grab the *outside front brake*, which slows down the rotation of the skid, allowing the vehicle to straighten and enabling you to regain control. In the case of Under Steer where the front tires have lost traction as you turn into a corner, the yaw sensor recognizes the car is still going straight but the steering wheel sensor says the wheel is turned, so there is definitely a problem. The system will again cut power and activate the A.B.S. which will grab the *inside rear brake* on the side toward which the car is trying to turn. This creates a drag on that side that pulls the vehicle into the turn. Race Instructor Lee Ezell uses the analogy of paddling a canoe. When you hold the paddle still in the water creating a drag as the canoe is gliding forward, it will turn the canoe to the same side as the dragging paddle.

When these systems were first introduced they were pretty rough, in some cases almost violent. So abrupt was the action in some systems, it would actually rip the steering wheel out of your hand. Since then, manufacturers have refined the systems to where you hardly notice them activating, and in some cases you wouldn't know it at all if it weren't for that tiny light that "pings" on the dashboard of certain models. Yeah, that "pinging" means slow the heck down! Either you're speeding or road conditions warrant slower speeds, which again means you are speeding.

Stability Control is so effective it has become as common as seat belts. In 2007, NHTSA mandated that all vehicles be equipped with ESC - Electronic Stability Control - as a standard feature by the 2012 model year.#3

Stability Control however is not an end-all-be-all. Its effectiveness is still governed by the Laws of Physics and the traction capabilities of your tires. You drive off a cliff you're not going to make a U-turn.

I'd like to point out, once you begin to master the Secrets of a Race Car Driver you will be able to recognize and correct a skid before the Stability Control intervenes.

Probably one of the most ingenious Stability Control Systems is the one introduced by Volvo that is designed to stop the rollover of SUVs. Okay, I was an instructor on their Master Safety Program at one time but again, no, I do not work for Volvo. They just happened to have introduced a number of the safety systems we take for granted today including the 3-point seat belt you're wearing every day. You are wearing it, right?

Rollover has been a serious issue ever since the introduction of SUVs. Crossed over from agricultural vehicles after World War II, SUVs - Sports Utility Vehicles - became very popular because you could take them anywhere, off-road or on. Today, of course, most SUVs never see dirt but people do enjoy the "captain's chair" view they have of the road. But this higher view is what creates the problem most associated with SUVs: you have a higher Center of Gravity (CG) and a greater chance of rolling the vehicle over.

The CG in a racecar is generally at or below the axle line - an imaginary line drawn between the centers of your wheels. In the average sedan the CG is just above the axle line. In an SUV, it is generally a foot or more above the axle line. This higher Center of Gravity can cause the vehicle to tip or roll over more easily.

Some of the first SUVs were so narrow the vehicles could roll on the first attempt at an abrupt turn at speed. Today most SUVs are wide enough, that type of rollover is generally not an issue. However with today's vehicles it is not the first swerve that flips an SUV. It's the *second* swerve when you are trying to straighten the vehicle.

Here's how rollover works and it's based on one of the Fundamental Laws of Physics: energy cannot be created or destroyed, it only transforms. Let's say you swerve to the left in an SUV at highway speed to avoid a car that has slammed on its brakes in front of you. As you swerve left, all the energy that was propelling the car forward is now going into the springs on the right side of the vehicle causing them to collapse. Because SUVs sit higher off the ground, this abrupt compression of the springs is storing an inordinate amount of energy into those springs. As you swerve back to the right to avoid possible head-on traffic, lateral G-force will toss the weight of the car's body off the springs before the springs have a chance to release. When the springs do release, the action is so sudden, they will push the weight of the car's body past the tipping point, launching the vehicle over in much the same way the lid on a Jack-in-the-box is popped open by the spring inside.

The Anti-Roll System is designed to recognize when there's too much energy in the springs and triggers the System to intervene. This is Stability Control on steroids. It grabs a combination of front and rear brakes. So before the second swerve begins when you attempt to straighten the car, the Stability Control is changing the pivot point of the vehicle from the front to the rear. The result is the front end comes around rather than the rear end, effectively causing the SUV to flop down and not allow a roll over…brilliant engineering, to say the least.

But again remember, the Laws of Physics will come into play; they will *always* come into play. If you hit something while sliding sideways, a curb, a pothole, you're going over. Not even a racecar can keep all four tires on the ground when it hits something going sideways.

THE GLAMOUR OF RACING

It was a race weekend with my Formula Car. My buddy, Dave Carman (yeah, it's an ironic name), came with me to be my mechanic and wrench on the car. We had a deal to trade weekends: he wrenched for me when I was racing, and I wrenched for him when he was racing his dirt track Midget racecar.

In late qualifying on Saturday afternoon, I blew the engine. There were a lot of weird knocking sounds as I quickly shut the engine off and coasted off the track - watching dollar bills fly out of my wallet. It doesn't take much money to race. It just takes all the money you've got.

We pulled the engine out of the car and took it to a fellow racer's machine shop where we spent most of the night taking it apart, replacing the broken stuff and putting it all back together again. We arrived back at the track and finished putting the engine in the car sometime after three in the morning. We managed about three hours sleep, got up, threw down a coffee, spent the rest of the morning realigning the car and getting it prepped with "nut and bolting" to make sure everything was tight. Adrenalin was taking care of the fact I was hot, sweaty and tired when I slid into my fireproof race suit. We pushed the car up to the starting grid for the first race. I was putting on my helmet when I looked up to see people hurrying into the stands. There was a crowd of people pressed against the fence at the edge of the track, taking pictures, families with kids their eyes as big as silver dollars. You could feel the excitement in the air.

I took a moment to reflect on what was happening. I began to laugh, just shaking my head. I had forgotten…this was glamorous.

THREE WAYS TO LOOK AT CAR SAFETY

What about the Secrets?? Relax. I'm going to get to the Secrets shortly. First there are some things you need to know about car safety, and what you need to look for when you are either buying a car or taking a closer look at the car you have now.

The first way to look at car safety is what is often referred to as PASSIVE SAFETY. This is everything a car can do to keep you alive once you are *in* a collision. While you are impacting some object at high or low speed, a number of engineering devices are working to keep you safe or, in worse case scenarios, to keep you above ground for another day.

The most obvious of these devices are the Airbags. You can't have enough Airbags. The more the merrier. If I'm in a collision, I would like the whole interior of the car to turn into one large balloon that just bounces me off everything until the dust settles. Airbags, however, are not designed to bounce you but rather catch you and slow your progress toward the stuff that is definitely not soft.

The Federal government required Front Airbags by 1999 but most manufacturers had Driver Airbags by 1995 and Passenger Airbags by 1997. #4

Front Airbags are generally duel or multi-staged Airbags that inflate to the necessary pressure for the severity of the impact. But not all front Airbags are duel or multi-staged. Check with your dealer or, more specifically, the technicians who replace the airbags. Also check the Maroney: it's the window sticker. Every new car is legally bound to have one. It lists all the features in that specific vehicle. Check it to find out what Airbags are in the vehicle.

The Airbags are designed to work in conjunction with your seatbelts and proper seating position. You want your breastbone to be at least ten inches from the steering wheel. But not so far back your breastbone is a couple feet from the steering wheel.

Remember, the Airbag is designed to *catch* you and slow your progress. And it happens faster than you can blink. It's a pyrotechnic explosion that happens in 1/25th of a second at up to 200 miles per hour.#5 The bags are then designed to deflate through vents. If you are sitting too far back, the Airbag could already be deflated by the time you get to it, defeating the purpose of the Airbag. And yes, that impact is going to make an impression. Too close to the Airbag and it could crack your sternum or worse. For the petite or shorter person, you want to find a vehicle with enough seat adjustment and, if available, pedal adjustment that enables you to find that ten-inch minimum distance from the steering wheel. A more recent innovation in some models is the Smart Airbag: it determines pressure based on the weight of the occupant on the seat.#6

There are also Side and Curtain Airbags designed to protect you from side impact. They, along with your seat belts, protect your torso and keep your head from going out the side window, which during a second impact could prove fatal. They also help to protect you from flying glass.

Knee bags are designed to keep you from submarining your seat belts. It comes in handy for all those Darwin Award contestants who lean their seat so far back their head is barely above the windowsill. I remember an "accident" where a guy in a compact slammed into a parked car. His seat was laid way back, and on impact he went under his seat belts and broke his legs when his kneecaps impacted the dashboard.

For Child Safety I would like you to consider this: an adult would impact an Airbag predominantly with their upper torso and head, which is how Airbags are designed. A child would be impacting an Airbag with just their head, in most cases killing them.

I cannot emphasize enough how important it is to have infants and children seated in the back seat with proper supplemental seating:

infants in rear-facing baby seats, and children in booster seats. This puts them in a better position to utilize seat belts and Airbags properly. Do NOT put an infant in the front seat. I know, they're a newborn and you just can't take your eyes off them. Those are two reasons not to put them in the front. First, you'll be distracted while driving but more importantly if you have a collision the Airbag will be hitting the baby seat at up to 200 miles per hour. Which do you want: a healthy baby or some lame-after-the-fact lawsuit against the manufacturer because the front passenger airbag switch malfunctioned? There are states that forbid the baby seat in the front, but beside the law, do you really want to take that risk? Check the Manufacturer's recommendations or go to NHTSA's website for recommended seating for infants and children: www.**nhtsa**.gov/ChildSafety/Guidance.

States have different requirements but check the Owner's Manual for the Manufacturer's requirements. It's not just the child's age but more importantly there are height requirements so the seat belts are being properly utilized, positioning the belt across the torso where the rib cage protects the internal organs as it would with an adult, and not across the child's neck. Small children will often flip the torso belt behind them rather than have it across their neck. A lap belt by itself in a collision can cause possible internal injury.

Get a booster seat! Again, do you really want to risk your child's life?

Passive Safety also includes door braces to protect from side impact and roofs strong enough to hold the weight of the vehicle. The better roofs are rated at supporting four and five times the weight of the vehicle. A car is usually airborne before it lands on the roof, so the distance of the drop before landing multiplies the weight (LOAD, for you physics majors).

I want to take a moment to talk about crush panels. A great deal of study has gone into understanding Collision Dynamics - what the heck is going on when two vehicles collide or when a single vehicle hits an immovable object. We've all seen the crash dummy commercials. Again, it's all about energy and the Laws of Physics. Energy cannot be created or destroyed; it simply transforms. When you hit something, the ENERGY that was your SPEED is now going to TRANSFORM into and through your vehicle and it's going to start with the softest material it can find - YOU.

This fact was brought home tragically when one of the great racing icons, Dale Earnhardt, was killed at Daytona. When his race car hit the wall on the last lap of the race, all the energy from that impact went into his neck killing him. Today, most vehicles are designed with crush panels, softer metals in the front and the back designed to collapse on impact thereby absorbing much of the energy in a collision. Racecars are even being designed with crush panels to help absorb that high-speed energy, one of the few instances where family car engineering has found its way into racing rather than the other way around.

Since Earnhardt's death almost all car racing requires some kind of head restraint system that supports the head and keeps it from snapping too far forward. That doesn't mean you have to immediately run out and buy a head restraint system. But if states keep giving driver's licenses without some better form of required training and proficiency, who knows? Pretty soon, we may be required to wear helmets and head restraints just to get to the grocery store. In the mean time, make sure the vehicle you buy has energy absorbing crush panels to suck up the speed energy when you can't avoid the collision.

The second form of car safety is called ACTIVE SAFETY - what the car can do to help you avoid the collision in the first place.

Who'd a guessed, but your seat belts are considered Active Safety because you Actively have to put them on. Yes, use them - the chances of them being a detriment are far less than the advantage they provide. Well, what if I land in water?? First off, you need your seat belt to survive the impact of hitting water. It can be like hitting cement; it can knock you out and you'll drown anyway.

Why do you think high divers only dive into aerated pools? It's to soften the water so they don't kill themselves on impact. As I was saying, yes, wear your seatbelts.

Another piece of automotive engineering that is part of Active Safety is how well your car handles. This is one of the lessons learned from European car manufacturers who were more closely tied to racing: a BETTER HANDLING car is a SAFER car. By better handling I mean a car that has responsive steering and very little body roll or lean when you turn the wheel.

Body roll is deadly and here's why: when a car leans too much going through a turn, all the weight (again, *load* for you physics majors) is going onto the outside tires and coming off the inside tires. At that point only your two outside tires have the grip while the inside tires are losing grip. This will destabilize the vehicle and cause it to skid. By keeping the car more level as it goes through a turn, you are balancing the weight of the vehicle over all four tires rather than just two. This gives the car more tire grip and therefore gives you, the driver, more control.

Keeping a car level in a turn is one of the fundamentals of racecar design. Racecar suspensions however are much too stiff for everyday driving with its bumps, potholes and high curbs. You would have to wear a kidney belt. Truck drivers used to wear them until someone was smart enough to put soft springs between the truck's cab and it's frame.

Every vehicle is a compromise, a balancing act of handling for safety and ride quality for comfort. There are vehicles that have electronically adjustable suspensions that can be dialed from soft to stiff but these vehicles are generally very expensive. For the rest of us who are just trying to get to work or the grocery store, we need a cheaper balance of handling and ride quality.

This is where sway bars come into play. They are part of a system that relies on a balance between softer springs for comfort but stiffer sway bars for handling. Sway bars take effect when you go into a corner. As the car starts to lean in the corner, the sway bar is designed to push down the opposite side of the car, in effect keeping the car level as you go through the corner. Once again, keeping the weight on *all four* tires makes for better traction and control.

Think of it this way: a car is nothing more than an extension of your arms and legs. Just as you are able to jump out of the way of danger with your legs, you need to be able to make your *car* jump out of the way of danger. To accomplish this, you have to know what your car is capable of doing. Otherwise you may not try to use it when you need it.

An easy way to find out your car's capabilities is to test your vehicle's body roll in a BIG EMPTY parking lot. Make sure your seat belt is good and tight or you'll start sliding around in your seat. You won't be able to tell what your car is doing because you'll be too busy hanging on to the steering wheel rather than controlling the steering wheel.

At SLOW SPEEDS (10-15 MPH – Keep it a *controlled* test.) move the steering wheel back and forth, at first gently and then gradually more aggressively as if avoiding a collision. You want to pay attention to how stable the car feels as you move it from side to side. There's going to be some lean, but does it lean too much where it feels unstable? Or does it feel solid under you, giving you confidence rather than feeling uneasy? The less body roll, the more stable the car will feel, the more control and confidence you will have when driving even with an aggressive emergency turn. If you own the car and it's flopping all over the place, you might want to talk to your mechanic about a heavier sway bar or new stiffer shock absorbers. Either that or it is time for a newer car.

WARNING: be prepared, that if the car flops suddenly from one side to the other, to get your foot off the gas pedal! This will allow the car to settle. More about this when we get to Secret #12 and a more in-depth explanation.

You also want to pay attention to how quickly the car responds to your steering input. There is a direct relationship between steering response and body roll. Once again it has to do with the Laws of Physics. A body in motion tends to stay in motion until acted upon by an outside source. In other words a car keeps going straight until it turns. I know on the surface this sounds stupid but here's what's happening: you turn the wheel, the front tires turn, BUT, the car body keeps going straight! What connects the body to the wheels is the suspension - springs, shock absorbers, a-arms and sway bars. As the body continues going straight, it pushes against the suspension causing the springs and shock absorbers to collapse or set. Once the suspension runs out of travel or sets, the body is forced to turn. You don't notice this because it happens in split seconds. But it is measurable. And it *is* a safety issue. The more body roll, or in this case the more dive or forward lean, the longer it takes for the suspension to set and the more distance you are covering before the car actually turns. If a kid runs out in front of you, the longer it takes for the suspension to set and for the car to turn, the closer you are getting to the kid! Less body roll means a quicker steering response and a better chance of avoiding tragedy.

A.B.S. and Stability Control are also part of the Active Safety capabilities of your car.

Cruise Control is a convenience application when you are on a wide-open highway and you can maintain a steady speed. But make sure you know how to shut it off, and practice doing so. You don't want to be surprised by a slower car and suddenly find yourself frantically searching for the OFF Button. The brake pedal stops it, but does not deactivate the system. When you go back to the gas pedal you may be surprised to find the car accelerating more than you want! No, it's not the dreaded Unintended Acceleration! It's the Cruise Control going back to your original speed setting. Just step on the brake again or push the Activation Button. On most systems the button that Activates Cruise Control also De-Activates it. Check your Owner's Manual to be sure.

Some manufacturers have introduced Active Radar Cruise Control Systems where you not only set the speed but also set the safe distance you want to keep from traffic ahead. As you come upon another vehicle and enter that safe distance, the system will begin slowing the car by retarding the gas pedal. Some systems will even apply the brakes but be aware these brake applications are generally no more than twenty percent. Most will NOT STOP the vehicle. Check with your dealer as to whether you have this system or not and know where the Activate/De-activate button is located. If you are driving in visually impaired conditions like rain or light fog, just know that on most vehicles your Radar Cruise Control will not operate below 25 -35 MPH.#7

There are systems that will STOP your car for you if you are not paying attention to the traffic ahead. DO NOT ASSUME YOUR CAR HAS THIS SYSTEM!! Find out from your dealer. This is a Warning for potential Darwin Award recipients (DAW).

There are Blind Spot Monitors that warn you of a vehicle at your rear corner panel in the next lane, and there are cars with Lane Alert/Keep Assist that when you are drowsy will warn you if you start to cross the lines on the road. There are also cars now that park themselves. There are even systems that recognize when you are not looking at traffic ahead, distracted by a billboard or *texting*, and if there is an impending problem it will "beep" to get you looking forward. Again, do not assume. If you're not sure of the safety systems in your car, ask your dealer and ask how to operate them.

The third way of looking at car safety?

As I said, some of the stuff car manufacturers are coming up with are down right amazing; but as marvelous as these machines are, they do not drive themselves. There are companies working on it. But until then, *you* drive the car. The car only does what *you* tell it to do. So the third and final way to look at car safety is YOU! The decisions *you* make and the actions *you* take behind the wheel determine whether *you* live or die.

And that brings us to the SECRETS OF A RACECAR DRIVER. These Secrets can help *you* make better decisions and take better actions to STAY ALIVE WHILE DRIVING.

TO FINISH FIRST

Today on the back straight at Watkins Glen Race Track there is a Bus-stop Chicane, a series of tight turns, just before the end of the straightaway to slow the cars down. It was put there after, most notably among others, Tommy Kendall had a horrific crash smashing head-on into the "blue bush" at high speed. To this day he still walks with a limp. Back when I was racing, before the Chicane was put in, it was a straight shot down the back straight heading into what was a daunting right-hand turn. It was a gut-check turn where you came over a slight rise at full speed and then dropped downhill as you went into the turn and came uphill as you exited the turn. Lining the outside edge of the road surface were turtle shells, little round mounds of asphalt that marked the edge. Beyond that was the "blue bush":

A BABY BLUE STEEL ARMCO BARRIER!

The first time I got to the end of the back straightaway and the right-hander at full speed, I "threw out the anchor" braking and downshifting. As I went through the turn I realized I was way over-compensating. The next lap I didn't downshift but I did brake. I still had a good two and a half feet to the edge of the track and the Armco barrier. Next time, I just lifted off the throttle as I dropped into the downhill portion of the turn but the rear end of the racecar was starting to slide out in Over Steer. On exiting the turn, however, I noticed the *uphill* portion was catching the rear of my car and straightening me. And I cleared the Armco by a foot and a half. So I figured I *should* be able to stay full throttle – gas pedal on the floorboard - and the uphill would still catch me and straighten the car just before getting to the Armco barrier.

I would have to drive the Line perfectly, using the entire width of the road creating the largest arc through the turn, to be able to take the turn without lifting off the throttle.

As I went careening down the back straight my right foot had the gas pedal jammed to the floor. It took all my strength and concentration to keep it there. It felt like every cell in my body was screaming LIFT! As I was approaching my turn-in point I looked over at my apex point, a spot just past the center of the inside edge of the turn. My peripheral vision caught my turn-in point and I cranked the wheel in one smooth motion. As the car dropped into the downhill, the rear of the car started to slide out in Over Steer and I had to correct my steering back to the left to keep the car turning right correctly toward the apex point. I looked ahead all the way to the exit of the turn. I shot past the apex point, my right front tire just touching the inside curb. I could see the "blue bush" in my peripheral vision to my left approaching as my car was sliding sideways under acceleration through the turn. I kept the gas pedal jammed to the floorboard – if I lifted now I would swap ends in a horrific Over Steer and would hit the Armco barrier head on. I kept looking down the track waiting, waiting, the car sliding sideways, the Armco barrier getting bigger and bigger in my peripheral vision. I felt my left rear tire skipping across the turtle shells like a drum roll! Then suddenly I was into the uphill portion of the turn, the car snapped straight and I continued on down the track.

After that, it was flat out - full throttle - every time.

A lot of people have the misconception that car racing is all about driving with basketballs. It's not a testosterone contest. Don't kid yourself. There are times when you *will* be tightening your seatbelts. But car racing is a thinking person's sport - discovering on and off the track what it takes to go as fast as possible and still remain under control. As you are searching for the "edge," that balancing point between ultimate speed and catastrophe, your nerve will be tested. But you have to approach it gradually, not like the last words of the redneck that yelled, "Hey y'all watch this!"#8

The way to approach racing was best stated by three-time World Driving Champion Nikki Lauda: "You don't take chances; you take calculated risks."

I can't tell you how many times I've been right-seat coaching to look over and see my student hunched over the wheel with a death grip on the wheel and their shoulders up somewhere around their ears. They've got the combined look of someone about to kick the crap out of somebody and the panicked look of someone going over a cliff. I've reached over and grabbed their shoulder muscles at the base of the neck and squeezed; usually getting a howl of pain followed by a "What the…?" Then I calmly explain they need to actually breathe while doing this. You'd be amazed at the number of people for whom this is news, completely unaware they were holding their breath.

Obviously guys are the worst, since they come with a testosterone bottle attached. But it amazes me how many people think it's about balls instead of brains. There will come a time when you will have to overcome fear if you're going to pass that guy in front of you. But that's only after you've figured out where the Line is, where the others are missing the Line and not getting off a corner fast enough. That's when you can pass them on the straight because you were on Line through the corner and carried more speed onto the straightaway. The guys that *win* races are setting up their competition for the pass, not making banzai moves that invariably end in carnage.

As 4-time Indy 500 winner Rick Mears said: "To finish first, you must first finish."

SECRET #1

(All you guys who skipped the beginning to get here – Go back – You're going to need it.)

I've been teaching racing for a number of years with race schools and race programs as well as coaching private clients. What I've found is that race schools and emergency driving schools teach you HOW CARS WORK and they teach you WHAT TO DO when the car does something scary like skidding or sliding. If you go to these schools they will definitely make a better driver out of you. If your car skids you have a better understanding of what to do, you'll have a greater awareness of your driving, and of course the schools are just loads of fun!

But very often as my friend Ron Fieni, a talented race driver and instructor, pointed out: what is taught in the race schools and the emergency driving schools doesn't always translate. When you are on the racetrack you are very aware the car might skid. At an emergency driving school doing the skid exercise, you *know* you are going to skid. At the crash avoidance exercise, you *know* you're going to have to avoid something.

But everyday driving isn't like that. When you go to the store, you're thinking about the store. You're not thinking about the kid who is about to run in front of you. You're not thinking about the patch of ice you're about to skid on.

What the race schools do *not* teach you is…HOW *YOU* WORK!

Remember when I was talking about all that knee-jerk (read: panic) reaction stuff? I talked about A.B.S. and how engineers recognized people slam on the brakes *in a panic* and turn the wheel at the same time. Well, what nobody tells you is why you do that. I know this sounds silly since the answer would seem obvious and simple.

But the answer is not as obvious or as simple as you may think….

I like to explain it this way: we have a common ancestor.

This is a guy or a gal a million years ago who looked at the saber-toothed tiger, saw the danger, and jumped. The guy who just stood there? Got eliminated from the gene pool.

So looking at danger and jumping have been handed down to us. It's part of our DNA. It's instinctive. We all have it. It's called "Fight or Flight."

If lightening hits near you? You're going to come off the ground and you'll be looking at where it struck, so strong is the fight or flight reaction. Someone jumps out and yells BOO! In a split second you're going to look right at them and jump.

As a pedestrian, fight or flight is great. A car comes at you out of control, you look at the danger and you jump out of the way. After all, fight or flight has been helping us survive for millennia.

But when you are behind the wheel driving a car? It is the worst possible thing you can do! That looking at danger and that jump reaction, you have *no control* over it! It is an *out-of-control* reaction! You are instinctively going to look at the kid or the tree or the other car coming at you. You instinctively are going to jump! Think of it this way: when you are behind the wheel of a car, you are in control of a weapon. So when you are behind the wheel of a car and fight or flight kicks in, you are:

OUT OF CONTROL with a WEAPON!

Once again, what is the last thing you see before you hit the tree? And don't say, the inside of your eyelids. I'll give you the Darwin Award now.

What is the last thing you see before you hit the tree??

…The tree!

The reason you hit the tree?

Because you were looking at it!

You're driving around a corner and for whatever reason the car starts to skid. Maybe you hit some gravel, or ice, or oil on the road. Couldn't have been you were going too fast... So now you're skidding sideways and you're skidding right toward the tree, and what is that fight-or-flight-knee-jerk reaction? Oh my god! I'm skidding towards a tree! You look at the danger! You look at the tree! The only problem is you are behind the wheel of a car. As soon as you LOOK at the tree, guess what?

You are now STEERING toward the tree! Where you LOOK is where you STEER! It is Eye-Hand Coordination. Every stick and ball sport and just about everything else we do is based on it. Your hands automatically follow your eyes.

The Secrets of a Racecar Driver – the specific techniques of racecar driving - are designed to help mitigate, to help stay one step ahead of the "fight or flight" reaction. And it starts with the one thing every race driver and *you* must *always* do and it is a Race Car Driver's:

SECRET #1

YOU ALWAYS LOOK **AHEAD** WHERE YOU **WANT** THE CAR TO GO.

Not where the car IS going, but where you WANT it to go! For racecar drivers, this means never taking their eyes off the track ahead. The car can be and often is skidding all over the place as they squeeze every bit of speed out it. But as long as they keep their eyes on the track ahead, their hands will automatically adjust the wheel to keep the car pointed at the track - where they want the car to go.

If you've ever seen in-car camera shots of race drivers in action, you'll see that on occasion they will jerk the steering wheel slightly in the opposite direction of the turn. It means they are correcting the steering as the car skids. If we had a simultaneous camera shot looking down on the race car, you would see the rear end of the car starting to slide in Over Steer. You would then see the driver jerk the wheel slightly to correct the steering and keep the car on its intended path. Sound familiar? It's what I described the Stability Control System does, keeps the car on its intended path. Race Drivers are adjusting the steering wheel automatically because they are looking ahead at the track.

There are three reasons why Secret #1 is so important - LOOKING *AHEAD* WHERE YOU *WANT* THE CAR TO GO.

As I mentioned, where you LOOK is where you STEER is the first reason, so we don't want to be looking at that tree.

The second is that by LOOKING AHEAD you will more quickly FEEL if the car is skidding.

The third reason, and this is the most important for protecting yourself against irresponsible drivers, when you LOOK AHEAD you ANTICIPATE what will be happening ahead.

Let's take the first one: Where you LOOK is where you STEER. Think of this as Secret #1A. When you skid, don't kid your self, fight or flight is going to kick in. Adrenalin is going to be dumping into your system to prepare you to jump and run. You *are* going to look at the "saber-toothed tiger," you *are* going to look at the tree. Millions of years of evolution are at work here. In the case of race drivers they are going to look at the wall or another racecar they are skidding toward. But race drivers train themselves to not freak out over it, to not panic because they know they have options. What they do and what *you* must learn to do is immediately recognize you are in fight-or-flight mode – you are looking at the danger. Now, you have to adjust your vision to look where you want the car to go. Where is the road? When you look at the road *ahead* your hands will automatically follow your eyes and point the steering wheel toward the road. When the tires regain their traction the car will continue down the road where you are looking.

Remember Driver's Ed. in high school when the driving instructor use to say, turn toward the skid? In other words, if the rear of the car skids to your left, you would turn the steering wheel to the left. This is partially true. You do have to adjust the steering toward the direction you are skidding but only to the point where you are steering toward the road, not toward the tree or as in the case of a race driver toward the wall.

In Figure 1 we see the rear tires skid - Over Steer - to the left in a right hand turn. We see the driver looking toward the skid - and the tree - and then over turning the steering wheel, sending the car right where he's looking…the tree!

Fig. 1 Fig. 2

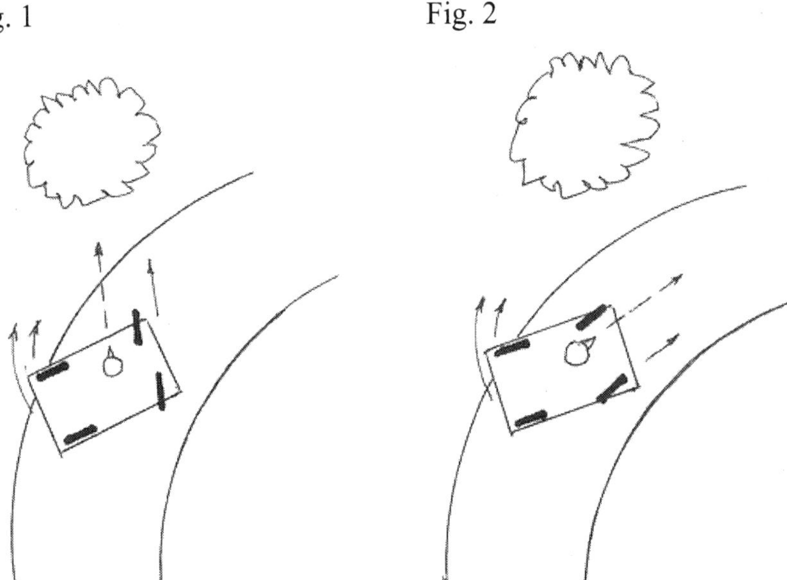

In Figure 2 when you keep your eyes on the road ahead you automatically turn the steering wheel to keep the car pointed toward the road. Once the skidding rear tires regain traction, they will fall in line with the front tires.

For those of you driving in snow in the winter think about how the car seems unstable, sliding from side to side. As long as you keep *looking* at the road ahead, your hands will automatically adjust the wheel to keep the car *pointed* at the road ahead even though the car is sliding around. Knowing that by looking ahead you can control the car helps you maintain control of yourself, staying one step ahead of the panic reaction of flight or fight.

Remember: keep looking at the road *ahead* where you *want* the car to go.

I was teaching students how to drift a Mercedes C63 at a Mercedes AMG program. A fun little car if you can afford it. I tossed the car sideways into the corner. (CLOSED COURSE, PROFESSIONAL DRIVER - DAW - again, a Warning for those thinking of trying for the Darwin Award). As the car slid into the turn, I glanced in the rear view mirror to see the students in the back seat all turn their heads to look where the car was sliding.

I asked if there was a naked girl over there.

They laughed but point taken.

HURRY UP, YOU'RE LATE

At the moment a car starts to skid, you are *late* with your steering. When the rear end skids in Over Steer, it will take a moment for you to realize the car is skidding. Initially you haven't moved the steering wheel, so in *that* moment the car is pointing toward the side of the road rather than down the road. We're talking split seconds here, but split seconds that can have serious consequences. If the tires were to suddenly regain traction the car could shoot right off the road! The car is steering you, you're not steering the car. So you have to play catch-up. This brings us to:

SECRET #2

WHEN STEERING in a SKID, USE **QUICK HANDS**.

Your adjustment of the steering wheel to point the car down the road has to be done quickly. You are playing catch up.

In Figure 3 we see the car has started to skid and the driver has not reacted yet. We then see the use of quick hands on the steering wheel to get the front tires pointed down the road, our intended path, rather than the side of the road.

Fig. 3

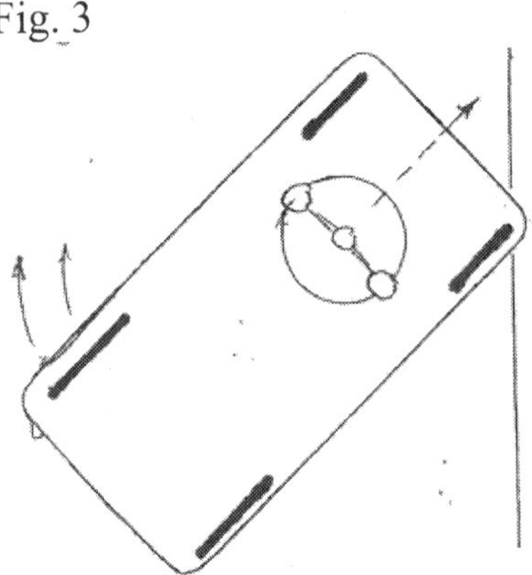

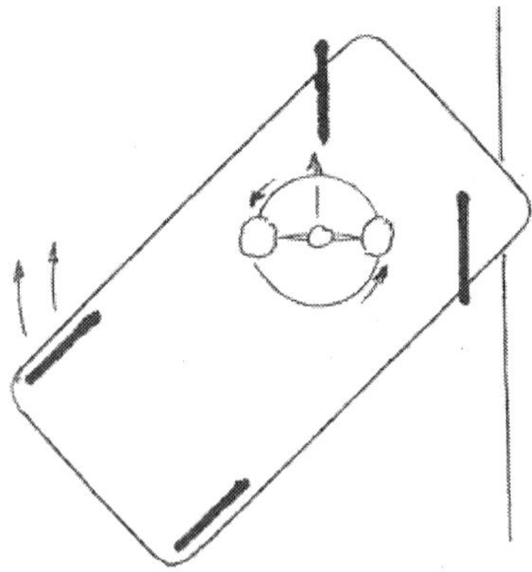

The same is true when the car starts to straighten from the skid. The rear tires will fall back in line with the front tires. So again, you must use quick hands to adjust the steering wheel back to keep it pointed down the road, our intended path.

Fig. 4

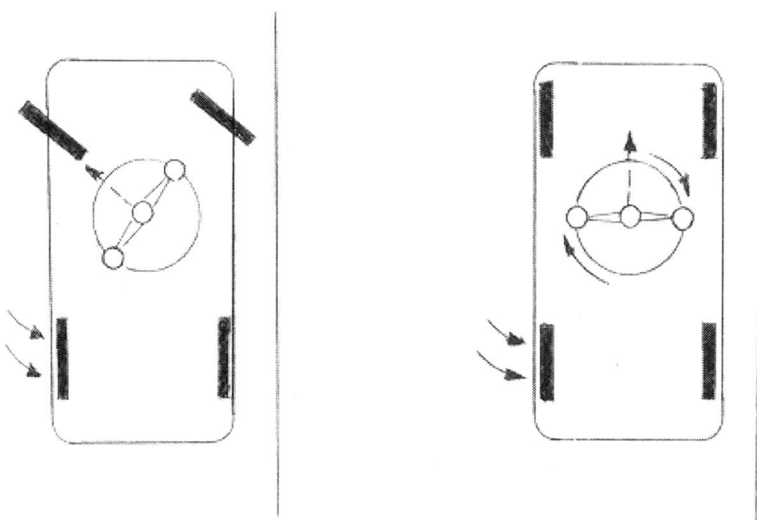

You turn quickly to CORRECT the steering as the car skids, and you straighten the wheel quickly to RECOVER the steering as the car straightens from the skid. The initial turn is Correction and the straightening turn of the wheel is Recovery. Some race schools will go further to describe the moment in between where the car is in a momentary state of equilibrium as Pause - Correct, Pause, Recover. Unless you're in a drifting contest, you will find the state of equilibrium or Pause is short-lived so be ready to straighten the steering wheel or Recover and do it quickly – Quick Hands.

SECRET #3

LOOKING AHEAD HELPS YOU **FEEL** WHEN THE CAR IS SKIDDING.

Vision is a very powerful sense. It not only can overpower your other senses it can even overpower common sense…need I mention texting?

If you ever rode a bicycle, when you first started you probably were looking down at the bike and it was shaking back and forth as you struggled to find your balance. When you Looked Ahead to see where you were going, the bike began to stop shaking because you allowed your body to FEEL the balancing point of the bike. Once you got more confident with Looking Ahead and Feeling the balance of the bike under you, it became second nature and you didn't have to think about it. Cars are no different. But for some reason when people get behind the wheel of a car their eyes are immediately drawn down to the hood. They are trying to keep track of the car visually. They are watching what the car is doing in relation to what is immediately in front of the car. This action gives the driver a false sense of security. This is particularly true of teens and I'll talk more about this later in the teen chapter.

Race schools have been known to tape the bottom half of the windshield to force the driver to look ahead. DO NOT DO THIS ON THE STREET (Darwin Award Warning)! When I'm coaching new race drivers I see them looking down just over the hood watching the car and what is immediately in front of it. I can tell when they're doing it because their hands are constantly second guessing how much to turn the steering wheel. As they turn into a corner, the students immediately start adjusting the steering wheel nervously, or what we refer to at the race schools as "sawing the wheel." They are looking directly over the hood trying to gauge the movement of the car visually, which invariably will lead them to react too late if the car skids either in Under Steer or Over Steer. I then have to force them to look ahead. I constantly point to where they should be looking but I've been known to reach over and hook my finger under their helmet and turn their head to where they should be looking and again point.

By Looking Ahead at the road rather than the front of the car, you take your eyes out of the process of reacting to the car. This allows your body to FEEL the vehicle and react more quickly if it is skidding - changing direction from your intended path. Grabbing the helmet may seem drastic but it is *that* important!

You don't look down at your legs and feet when you are walking. You look ahead and you FEEL your legs and your footing under you. Should your foot slip, you don't look down to correct it. You regain your balance by FEELING it. Yeah, you may look down afterwards to see what you tripped or slipped on, like "who put that there?" but your initial reaction does not use your eyes.

The same thing is true of a car. A car is nothing more than an extension of your legs allowing you to travel farther and faster. By Looking Ahead you will better FEEL the car and the tire grip under you. Just like your instinctive reaction when your foot slips, Looking Ahead allows you to respond quicker should your tires slip when they lose some of their grip – skidding.

By continuing to Look Ahead and Feeling the car under you, you will begin to establish a form of Motor Memory. You will start to trust it as you get used to Feeling the car and it will become automatic. This is where your confidence in Feeling what the car is doing will allow you to RELAX. It all goes hand in hand. The more you are able to RELAX, the more you will actually FEEL the car on the road. Your butt-end will start acting like a yaw sensor in a Stability Control System.

Ever hear the expression, "Fly by the seat of your pants?" Well in this case it is "Drive by the seat of your pants." And it is literally true.

To practice I recommend starting with a go-kart slick-track. Slick-tracks are usually concrete and powdered down with talc so they are slicker than the proverbial greased pig. The kart will be sliding all over the place so these tracks are ideal for practicing these techniques. If not available, *any* go-kart track will do since it doesn't take much to get a go-kart to skid. Keep Looking Where You Want To Go, Feel when the cart is skidding and use Quick Hands to Correct and Recover the skid! Practice, practice, practice!

As I said at the start, you can practice the Secrets in your everyday driving. If you were raised in snow country, you were probably taught from day one Skid Correction, Quick Hands, and Looking Ahead while Feeling the car. For those of you who are more snow deficient, RAIN and a BIG EMPTY parking lot will suffice. Again BIG and EMPTY and SLOW SPEED, you don't need to be going fast to get your rear tires to start skidding in the rain. SLOW SPEED! You also don't want to be tearing up your tires. Move the Steering Wheel around, get the car sliding. Don't panic, it's not the end of the world when the car starts to slide as long as you have plenty of room and you maintain a slow speed. When the vehicle starts to skid:

GET YOUR FEET OFF THE PEDALS!

Getting off the gas pedal and the brake pedal allows the tires to do *one* thing at a time, find traction so the car will settle.

Remember in a fight or *flight* reaction, the *flight* part is where your feet want to get active and jump. That's why it is important to keep the speeds slow so this remains a controlled exercise. Getting your feet off the pedals in a "real world" skid is not as easy as you might think. That's why practicing under these controlled conditions will help to establish a correct response when faced with the "real thing." I'll talk more about this in Secret #9 – *getting your feet to follow your hands*, and Secret #12 – *when in doubt, both feet out.*

Move the steering wheel back and forth until you begin to *feel* the rear tires skidding. It's just like the body roll exercise but we're taking it a little further, being a little more aggressive with the steering wheel to get the tires to slide on the wet road surface. Don't start looking at the hood of your car and trying to *see* what the car is doing. There is some visual recognition when the car is not going where you are looking, but keep Looking Ahead to where you want the car to go and focus on FEELING the car when it is skidding. As the rear tires skid, your hands will automatically steer the car where you are looking. Also remember to BREEEATHE – breathing helps to keep you relaxed which again helps you feel the car's movement.

Without taking too much tread off your tires, get a feel for what your car is like in a skid so when you are out there on the public streets and for what ever reason the car starts to skid, this won't be some new OMG experience! You'll recognize the universe is not upside down. This is just a momentary loss of traction and you have OPTIONS. You can take the steps necessary to correct this momentary loss of traction: Look Ahead where you Want the car to go, Breathe, Feel what the Car is doing, get your Feet Off the Pedals, Correct and Recover the Steering Quickly!

FIXING THE THREE REACTIONS TO FEAR

Remember the race-driving student hunched over with a death grip on the steering wheel? He was looking just in front of the hood with his teeth clenched together. When we look at the terms "fight or flight," what we see here is a perfect example of the FIGHT reaction to fear.

Also, I had a client who wanted to learn how to race his very valuable historic racecar without killing himself or his investment. We rented a track for the day and coming out of a particular corner that led onto the long front straightaway I told him to ease the gas pedal all the way down to full throttle. As we came out of the corner I could tell by the sound of the engine he was not at full throttle. As many times as we did the exercise he still would *not* go to full throttle - a perfect example of the FLIGHT reaction to fear.

Remember too, the students at the race schools who were nervously "sawing" the steering wheel as they turned into a corner? Instructors refer to this as "Low Eyes." They were nervously second-guessing how much steering to put in – a FLIGHT reaction - accompanied usually by being too abrupt with the pedals – a FIGHT reaction - causing the front end of the racecar to bounce up and down.

To fix this, we teach the students what is called "Eyes Up." We're not talking about looking up at the sky, we mean Look Ahead. In racing what is happening a quarter of a mile ahead on the track is going to be at the front bumper in a matter of seconds!

It always amazes my students when I tell them their fastest lap around the racetrack is going to feel like Sunday driving, as if they were just out cruising around the track. I know it sounds strange but it's true. The reason is because your fastest lap will come when you are relaxed behind the wheel. I'm not talking falling asleep behind the wheel although this has been known to happen to race drivers while they're sitting on the pre-race grid waiting for the signal to start engines – it's a reaction to adrenalin. I'm talking: focused, in control, and your body is relaxed. On the racetrack, that relaxation comes from what I consider *the* most important Secret of Race Driving:

SECRET #4

ALWAYS LOOK AHEAD TO **ANTICIPATE** WHAT IS GOING TO HAPPEN.

With my "fighting" student, remember I reached over and dug my thumb and forefinger into his shoulder muscles. He let out with a yell but I did it to show him he was completely locked up and looking no further than the front of the car. The first thing I needed him to do was to breathe. Next I needed him to stop looking just over the hood at the car directly in front of him and start looking as far down the track as he was able. Rather than having to be *reactive*, getting into a *fight* position for what he was afraid *might* happen, he became *proactive* when he Looked Ahead and was able to ANTICIPATE what was *actually* going to happen. He began to relax and breathe normally. He also stopped trying to take the leather cover off the steering wheel.

With my client who couldn't find the floorboard with the gas pedal, it occurred to me here's a guy who is CEO of his company and is used to being in control. Now he finds himself in a strange situation feeling out of control, not knowing what's going to happen, thus the *flight* reaction to fear. So I put an orange cone about a third of the way down the straightaway and I explained that as we came out of the corner I wanted him to Look Ahead to the orange cone and ease the gas pedal all the way to the floorboard. When we got to the orange cone I wanted him to lift off the gas pedal and begin slowing the racecar. Sure enough, as we came out of the corner I pointed to the orange cone and he eased the gas pedal all the way to full throttle, all the way to the floorboard. At full throat, Ferraris sound wonderful. When we got to the orange cone he came off the gas pedal and began braking and slowing the car. I moved the cone two thirds of the way down the straightaway. Again as we came out of the corner I pointed to the cone farther down the straightaway and he eased the gas pedal down all the way to full throttle. When he got to the cone he again lifted off the gas pedal and had to brake harder because of the higher speed. I moved the cone all the way to the end of the straightaway and the braking point for the next turn and told him this time brake really hard when we got to the cone. Again he eased onto full throttle and, carrying a ton of speed, braked hard at the orange cone slowing the car and was able to turn cleanly into the next corner. The light went on. Now he understood:

LOOKING AHEAD AND ANTICIPATING ALLOWS YOU TO KNOW WHAT IS GOING TO HAPPEN. KNOWING WHAT IS GOING TO HAPPEN GIVES YOU MORE CONTROL OVER WHAT WILL HAPPEN TO YOU.

Now he had control, now he could relax, and now he could enjoy the sport of racing. Of course after that, it was a matter of reining him in as he was going crazy with enthusiasm.

So let's apply Looking Ahead and Anticipating to your everyday driving.

The next time out on the road I want you to look as far down the road as possible. I know there are various rules about looking ahead – the three-second rule, looking five cars or as many as twelve cars ahead – but almost no one does them. Just look as far as you are able. And don't be deterred by that truck or SUV with the blacked-out windows that don't allow you to look through the car at what is happening ahead. Move to the edge of your lane enough to where you can look down the side of them to see what is ahead. DO NOT GO OUT OF YOUR LANE (DAW)! What do you see? Look as far ahead as you can. And look to the CENTER of your lane.

Someone once asked me to describe what racing was like. I said it's like threading a car-sized needle with only inches to spare at high speed. At first this may sound impossible, but if you *look* right down the middle of the lane you will *drive* right down the middle of the lane regardless of your speed. For those of you who wander in your lane or get too close to the edge of the road or the center lane, just look AHEAD to the CENTER of your lane and all that stuff will go away. If you really want to see if it works – use empty plastic garbage cans or half-full one-gallon water bottles and a BIG EMPTY Parking Lot and SLOW SPEED! Again, if you think you can fly, you jump off the curb not a ten-story building – SLOW SPEED. Look dead center between your plastic obstacles and that is exactly where you will drive.

In city driving you can see maybe one, two or three blocks ahead. Maybe you see a red traffic light and a stopped bus. What else do you see? I want you to get used to looking as far ahead as you can while driving. If you are on a highway maybe you can see a quarter of a mile down the road. I'm not talking about pinning your vision on a spot trying to read license plates a quarter mile away, this is called TUNNEL VISION and it blocks out everything around you. Just keep Looking Ahead. Move your vision around. Move it to what is closer to you. That car entering the highway and merging into your lane; is he slowing or is he on a collision course with you? It's an optical illusion, but vehicles on a collision course can appear to be standing still. You have to look directly at the vehicle to judge by the background if he is slowing or speeding up. Be careful not to steer toward him. If he appears to be standing still would indicate he is on a collision course with you and not slowing! Get ready to take evasive action. Who is behind you? Check your rear-view mirror. Who is along side your car? Check your side-view mirrors. Where along side? Next to you or back by your rear corner panel?

A quick note here about blind spots. I know people complain about not being able to see what is behind them because of a too-large C or D Pillar at the back of their car that "creates a blind spot." The A Pillar is what holds your front windshield, the B Pillar is by your head, the C Pillar is by the back-seat passenger's head, and in SUVs and Mini-Vans the D Pillar is at the back. That too-large C or D Pillar is there to protect in case of rollover. That pillar however does not "create a blind spot." The blind spot is created when the side-view mirrors are adjusted incorrectly. If you can see the side of your car in your side-view mirrors, your mirrors are adjusted *incorrectly*. You have created a blind spot at your rear quarter panels next to your rear tires. A car can be next to you and you will never know it. That's why you feel compelled to look over your shoulder to that pillar that's blocking your view. I know it's very reassuring to keep checking the side of your car to see if it's still there. But, if you are willing to give that up and adjust your side mirrors correctly, you won't have to look over your shoulder, you'll be able to see all the way around the car just by using your mirrors.

Here's how to adjust your mirrors so you don't create a blind spot: lean your head to the left almost touching your window and adjust your left side-view mirror so you can just barely see the outside door handle. When you straighten up, you won't see the side of the car. Lean your head to the right and do the same, adjust the right side-view mirror so you can just barely see the door handle. When you straighten up, you won't see the side of the car. Now you have removed the blind spot. To test it, drive slowly past a parked car. As the parked car leaves your peripheral vision it should immediately show up in your side-view mirror. If it doesn't, adjust the mirror out a teensy-bit more so you catch the first hint of the rear of the parked car. When the parked car leaves the side-view mirror, it will immediately show up in your rear-view mirror above your dashboard. This way you have a three hundred and sixty degree view all the way around your car. No blind spots. It takes a little getting used to. You'll still lean your head to make sure the side of your car is still there. And you'll probably still look over your shoulder. But once you begin trusting this new view of the road you'll find you can't do without it.

Like I said, you want to keep your eyes moving around. I'm not talking dizzying movement, common sense here. Most of the time you want to be Looking Ahead, but an occasional glance in the rear-view and side-view mirrors helps you keep track of who and what is around you.

ANTICIPATE, ANTICIPATE, ANTICIPATE!
Are those brake lights a half-mile ahead? Lift your foot off the gas pedal. If there appears to be trouble ahead, do you have a clear lane or shoulder of the road to move on to? You are anticipating what may or may not happen. We may not have saber-toothed tigers today, but we do have drunken drivers and people texting behind the wheel.

Again, the more you are aware of what is going on ahead and around you, the more control you have over the outcome of what is going to happen to you!

Race Drivers constantly anticipate what is happening ahead on the track. Someone once asked me how fast I was going in my racecar? My answer is as fast as the car will go. There are no speedometers in a racecar, only tachometers to tell you the speed of the engine and when to shift to the next gear. Some racecars just have shift lights that blink on telling you when to shift. There are other gauges and adjustable knobs but none of them tell you how fast you are going. As I'm coming out of a corner I'm trying to ease the gas pedal to the floorboard as soon as possible and still keep the car on the track. Once on the straightaway, I keep the gas pedal on the floorboard until the next corner. I have no idea how fast I'm going because I'm not looking at my speed. I'm too busy ANTICIPATING the location of my next braking point - where I'm going to slow the car down to get through and out of the next corner. I am also ANTICIPATING what the other drivers ahead and behind me are going to do. I am trying to figure out what their flaws are, either how badly they are driving the Line or how badly their car is handling the track by Under Steering or Over Steering.

In racing it is pretty simple to ANTICIPATE what the other drivers are going to do. We are all headed in the same direction and we all *want* the same thing.

On the public streets and highways?

God only knows what drivers want and might be baffled as well. There are cars moving all over the place and there is no way of knowing what those drivers are thinking or what they're going to do next. Half of the time, it seems *they* don't know either. It's a toss up, and at highway speeds it can get down right scary!

Any race driver will tell you exactly what I discovered when I began racing. You realize just how dangerous the public streets *really are*. You feel a lot safer on a racetrack.

WATKINS GLEN AGAIN

At the end of the front straight at Watkins Glen Race Track you go slightly downhill before you dive into Turn One a ninety-degree right-hander. With a slight bit of banking on the exit you are able to power out of the turn with some speed. From there it's a downhill charge through the gears getting to third or fourth gear just before the track bottoms out into what are called the Esses. They are a sweeping right-hand turn that immediately transitions to a sweeping left-hand turn, before you go uphill and out onto the back straightaway.

I should take a moment here to differentiate between what is a straightaway when describing a racetrack and what is a straightaway when *driving* a racetrack. As you would expect, when describing a racetrack, a straightaway is a straight piece of road where you don't have to turn the steering wheel. When *driving*, however, a straightaway is a section of track defined by the gas pedal. When I can get the gas pedal to the floorboard and keep it there that is the beginning of a straightaway. It doesn't matter if I'm turning, going uphill or down, as long as the gas pedal is still on the floor, that's a straightaway. When I lift off the gas pedal that is the end of the straightaway. In a way, racing is all about trying to straighten the racetrack out as much as possible: get the gas pedal to the floorboard as early as possible and don't lift until it is absolutely necessary.

Since the Esses at Watkins Glen lead onto the back straightaway, an actual straight piece of track, it's imperative to try to carry as much speed through the Esses as possible.

Two years after Francois Cevert was killed in the Esses during a Formula One practice session, a chicane was put in the middle of the Esses to slow the cars down. A chicane is a set of curbs that force sharp turns – right-left or left-right – thus slowing the cars before proceeding to the rest of the track. Similar to what you'd find at a street construction site where you had to go around barriers to get to the rest of the street. Only these barriers were curbing which happened to still be there when I started racing at the Glen.

I did not race in Formula One but in one of the smaller Formula classes where the overall width of the racecar was narrower. I discovered that the top three drivers in my class were able to get through the chicane without lifting off the gas pedal, keeping their engines in the power curve to maintain momentum, in other words flat out. Along with figuring out the rest of the track, I had to figure out how they were getting through the chicane without lifting off the gas pedal.

As you came down the hill out of Turn One and entered the Esses, the sweeping right-hand turn became a *sharp* right-hand turn as you went around the curbing they had added to form the chicane. That led to an immediate *sharp* left-hand turn as you went around the curbing on the left, and that blended into the sweeping left-hander as you exited the Esses and went up the hill to the back straightaway. I figured the initial *sharp* right-hand turn to get into the chicane would scrub enough speed to get through without lifting off the gas pedal. But, as I was about to find out, there was more to it than that.

I came down the hill at full throttle and entered the right-hand sweeper of the Esses. As I came out of the right-hand sweeper heading for the chicane, I had to "threshold" my brakes - just short of lockup - to keep from destroying my racecar. The chicane was coming at me way too fast to react to it accurately. I was lucky to get it slowed enough to get through with just a couple rough bounces off the curbing. I discovered it was physically impossible to look at the right-hand sweeper and then look over at the right-hand curbing of the chicane and still be able to react in time. Ninja and the Matrix movies aside, it would be the equivalent of trying to react to an arrow coming at you head on.

That's when it clicked. I figured out how the front-runners did it, the chicane of course, not the arrow (DAW).

I came down the hill out of Turn One full throttle accelerating through the gears. As I got closer to the right-hand sweeper and the entry to the Esses, instead of looking at the right-hand sweeper I looked over to the chicane and the entry curbing on the right side. What I figured out was correct – I could still see the right-hand sweeper in my *Peripheral Vision* on my left. Without lifting off the gas pedal I drove through the right-hand sweeper using my Peripheral Vision all the while focusing on the chicane curbing ahead. I came out of the sweeper looking directly at the chicane curbing on the right. As soon as my right front tire touched the inside edge of the curbing, I flicked the steering wheel to the left, the racecar slid through the chicane, and I continued full throttle through the left-hand sweeper up the hill and out onto the back straightaway. It was full throttle, flat out, every time after that.

Until….

But that's a story for later.

SECRET #5

Let's face it: it's crazy out there. Danger can seem to come from anywhere. A car on the highway suddenly swerves in front of you; or the guy next to you is wandering in his lane as he looks down in his lap texting. On a city street, you only need to be distracted for one second and a pedestrian or car is suddenly in front of you from seemingly out of nowhere. Trying to see every possible danger on a highway or a city street seems impossible.

…Or is it?

With our normal vision we look *at* things. I know, right now you're saying, uh-duh. What I'm saying here is when we look directly *at* something we tend to pinpoint our vision, filtering out the periphery. That may be good when you are doing brain surgery or picking up a hot coffee filled to the brim, but when you are behind the wheel it is the worst thing you can do. Behind the wheel there are so many things to keep track of and look *at*: other cars, road signs, pedestrians, not counting the distractions of billboards, the sights, or some attractive pedestrian. Trying to keep an eye out for the sudden movement of a car changing lanes or pulling out in front of you - how can you keep track of it all without being scared to death or going crazy?

As it turns out, there *is* a way to keep track of it all. And yes without being scared or going crazy. And it is actually quite easy. Not only that, you do it all the time; you are just not aware of it. It's called your *Peripheral Vision.* And that brings us to Race Driving's

SECRET #5

WHEN LOOKING AHEAD,

USE YOUR **PERIPHERAL VISION**.

What is Peripheral Vision? Look straight ahead. Stretch your arms straight out to your sides and wiggle your fingers. While looking straight ahead, bring your arms back until you can't see your wiggling fingers. Now, move your arms forward until you can just barely see your wiggling fingers, all the while still looking straight ahead. Hold that position. Turn your head from side to side and see where your arms and hands are positioned. Your hands should be slightly less than ninety degrees from your head position when looking straight ahead. If it's drastically less than that, see your eye doctor. That view of everything else from wiggling fingers to wiggling fingers when you are looking straight ahead is Peripheral Vision.

Peripheral Vision is one of those things we don't think about. It's there; we just don't think about it. When you reach across the desk for a pen, you don't think about everything else on the desk; you're just focusing on the pen. But if you *did* think about it, you would see everything else on the desk without having to look directly at each item. Try it. Look at something in front of you. Without taking your eyes off that object allow your vision to take in everything else. What you will notice is your focus will expand. It's what I call WIDENING YOUR VISION. You won't be concentrating solely on the particular object you are looking at; your focus and concentration will expand to include the objects around it. If something moves, say a tree branch blown by the wind outside your window, your Peripheral Vision would immediately pick up that movement.

At one of the race programs I was involved in, Ron and I developed a slalom exercise with a twist. Slaloms were a common exercise and you'd see them used in car-testing commercials on TV. You had to weave your way around cones that were set up at intervals in a straight line. Think of it as the movements of a slithering snake.

You would go to the right of the first cone and then to the left of the next cone then back again to the right as in Figure 5 below.

Fig. 5

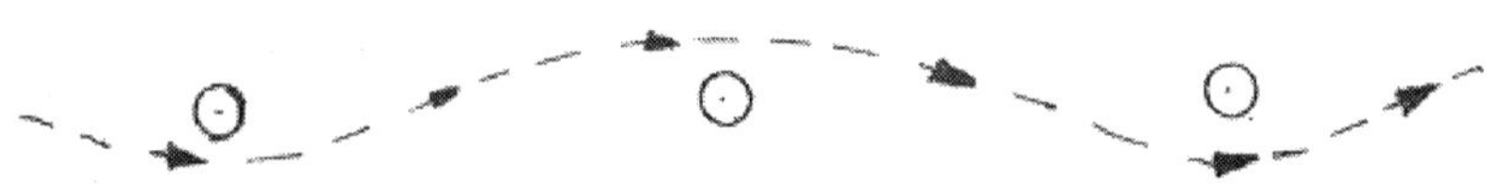

To get the participants to use and trust their peripheral vision, we explained that if they looked at the cones they would do one of two things: hit them since "where you look is where you steer" as in Figure 6. Or else they would drastically overdrive the course in an exaggerated movement to avoid the cones as in Figure 7.

Fig. 6

Fig. 7

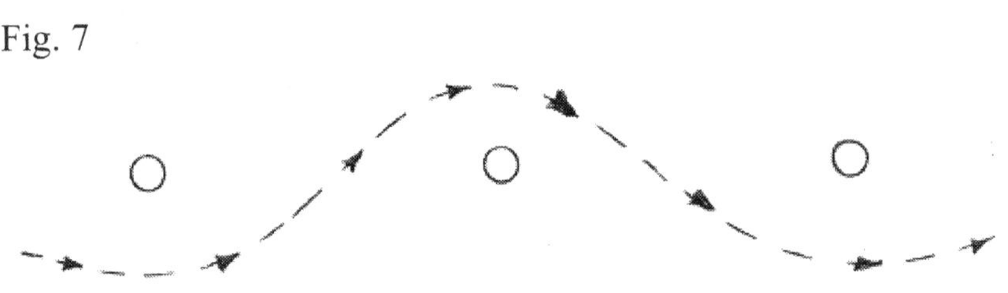

The "twist" we developed for the program was adding a big painted EYE on a pole set up beyond the last slalom cones and in line with them. We explained if the participants stopped looking at the cones and instead looked at the EYE that was just past the cones - Look Ahead where you Want the car to go - it would force them to use their Peripheral Vision and they would slither right past the slalom cones without hitting them.

So powerful was the fight or flight reaction to look at the danger that many hit the cones or else overdrove the course. We then slowed them down. Yes, it was a race school but you first have to get the technique down. By trying to race through the course, speed was just a distraction that kept them from working on the technique. Once they began looking past the cones at the big EYE and began trusting and using their Peripheral Vision, it was like an epiphany. OMG was a common response. The students slithered through the course without hitting any of the slalom cones and eventually with their fastest times.

When I'm in a race I'm looking as far ahead as possible down the racetrack to where the Line is - the most efficient path around the course. My peripheral vision will pick up the cars in front of me, and also tell me where those cars are off the Line so I can set them up for a pass. At the race school I can always tell when a student is looking at the car in front of him: whatever mistake *that* car makes, *he* will make the same mistake.

All right, let's take this out on the street. As you Look Ahead, I want you to practice relaxing your vision looking not at any one particular point. Again don't stare or fixate on some particular point, like the car in front of you, Look Ahead as far down the road as you can, relax your vision and allow your eyes to see everything, taking in as much as you can. Again, you are WIDENING YOUR VISION. This at first will seem strange and maybe feel a little spacey. Just practice it a little at a time. You may find yourself going in and out of this: widening your vision and seeing everything ahead of you and then pinpointing your vision to look in your rear view mirror or side mirrors…or steal a quick look at the car in front of you.

As you continue to practice Looking Ahead and using your Peripheral Vision to widen your vision and take in all that is in front of you, you will begin developing a unique skill. This is the ability to use your Peripheral Vision to see PATTERNS in the overall FLOW OF TRAFFIC. All the cars in your lane whether single or multiple lanes are all headed in the same direction.

You know this, of course, but knowing it and being aware of it using your Peripheral Vision are two different things. There is a flow of traffic moving with you going in the same direction and there is a flow of traffic moving in the opposite direction to your left - to your right in England, Japan, China, etc. A Pattern is formed. Once you are aware of this Pattern and see the flow with your widened Peripheral Vision, you will IMMEDIATLEY PICK OUT ANYTHING THAT BREAKS THE PATTERN of this flow. Like the branch blowing in the wind out of the corner of your eye, Peripheral Vision PICKS UP MOVEMENT or any break in the Pattern. It's that change in the status quo or the visual Pattern that Peripheral Vision picks up immediately.

Come to think of it, Peripheral Vision was probably also passed down by our common ancestor as a means of keeping an eye out for any unusual movement, like that saber-toothed tiger.

That car in front of you? You don't have to be looking directly *at* the car in front of you to see it. As you Look Ahead as far as you are able, your Peripheral Vision will easily pick up the car in front of you and anything it might do. You'll see if he's slowing, because as you get closer to him this movement or change in the status quo is a Break in the Pattern. You'll certainly see if his brake lights come on.

A car comes in from the right, a truck from the left, a pedestrian is starting to cross the flow of traffic, anything that breaks the pattern, your Peripheral Vision immediately picks it up. Whether it's vehicles from side streets, pedestrians, any movement at all, your *widened* Peripheral Vision will see it. On the highway there are brake lights a quarter of a mile ahead. Your Peripheral Vision picks it up because it's a Break in the Pattern. This is where you use your ANTICIPATION: you lift your foot off the gas pedal just in case. In the meantime, that SUV in front of you pulls away from you, oblivious, still speeding along - he's busy texting - until he gets to the stopped traffic and slams on his brakes!

In another scenario someone changes lanes, your Peripheral Vision sees it. You don't have to be looking directly at them to see it. However, the normal reaction *is* to look directly at the change in the pattern, to look directly at the car changing lanes. That too is part of fight or flight: is there danger there? Which is fine, don't fight it use it. It could be the car has blown a tire and is swerving out of control. Now, where do you look?? You Look Ahead where you Want your car to go – down the road! When you Look there you will Steer there. Your Peripheral Vision will keep track of where the swerving car is going and whether you have to take further evasive action!

You're driving down a residential street; you're almost home. You're Looking Ahead. The Pattern in front of you is a quiet street, cars parked on both sides. Your Peripheral Vision picks up movement, something breaking the Pattern. You look - fight or flight reaction - and see it is a bouncing ball. Is there danger there? In all likelihood, there's a kid or a dog chasing it. You ANTICIPATE what will happen next and you take your foot off the gas pedal and move it over to the brake pedal to be ready for a sudden stop.

Peripheral Vision is everything else you are seeing beside the particular focal point you've chosen. By Widening your Vision and seeing the Pattern in the Flow of Traffic, Peripheral Vision allows you to see more of what is happening ahead of you. And as you get use to using it and trusting it you will begin to realize just how powerful and effective Peripheral Vision can be.

When people are afraid of driving, and the highways can be particularly scary because of the higher speeds, it is because they feel they have no control over what might happen. Add to that, they often fear the worst. By Looking Ahead, Anticipating and using your Peripheral Vision enabling you to see more, you build a Cushion of Safety necessary to drive with control, confidence and the chance to get to the finish line. Okay, a little corny but I had to throw it in there.

I did mention there was a third reaction to fear. It's known as Freezing. In it's less extreme but more common form it's referred to as a "deer in the headlights." In racing we call it Target Fixation.

One of my first races was in a downpour at Summit Point Raceway in West Virginia. I bought my first set of rain tires. The tread on them was like a gumball; you could wiggle the tire tread like soft chewing gum. I had qualified mid-pack and since Formula cars don't have fenders when the green flag dropped it was just a wall of water in front of me, the spray from the tires. Two cars were side by side and all I could see were the little red "rain lights" mounted at the end of their transaxles. I looked off to the side to see if I recognized any markers that would let me know when Turn One was coming. All I saw were grass hills and trees flying past.

Suddenly I saw a car going to the right of the wall of water. It was Turn One. I braked and down-shifted to one gear higher than normal, a trick used in rain racing. I turned in and, as I straightened the wheel coming out of the turn, I squeezed the power on heading for Turn Two known as Wagon Bend. I followed the wall of water at full speed as we bounced up over a small hill in the middle of Wagon Bend. All of a sudden the wall of water parted, and there sat one of the front-running racecars dead center sideways in the middle of the track. To this day I can still see him. If I had frozen and Target Fixated I would have center punched him and quite possibly killed him. I looked to my left, saw daylight and yanked the wheel. In a blink I was around him. After the race I got out and kissed my rain tires.

Target Fixation or being "a deer in the headlights" can happen to anyone. Something suddenly happens in front of you! You start to stare at the danger! To help prevent Target Fixation happening to you, remind yourself: where you Look is where you Steer. So, Look Ahead where you Want the car To Go, and always practice SECRET #6…

WHERE'S THE EXIT?

On large oval racetracks, a race driver usually has a Spotter, someone high in the grandstand that is on their radio letting the driver know where the other cars are, and in the case of an accident letting him or her know where the track is clear to avoid the accident. You and I however do not have that luxury. And that brings us to:

SECRET #6

ALWAYS HAVE AN **ESCAPE ROUTE**.

This means keep your eyes moving and check your mirrors often. You want to know who's behind you or at your rear quarter panels at all times. If you have to make a quick maneuver to avoid a kid, you need to know where you have room; you need to know where your Escape Route is. In the case of crash avoidance when a car is coming at you suddenly whether from the side or head-on, you look for your Escape Route, daylight, any open space at all.

For example, you're pulling onto a busy road and you misjudge an on-coming car's speed – just pull onto the shoulder or as much of it that is available and get out of their way.

The light turns green. You enter the intersection. In your Peripheral Vision you catch a car running the red light, coming at you broadside. You have only split seconds to decide: where is your Escape Route? It might be a simple case of accelerating like heck to get out of their way. Or, can you stop in time? And if you can't avoid them, how can you *minimize the danger* to yourself and your loved ones with you? If he's going to side-impact you at the back half of your car, you might be able to turn slightly toward him so he hits the rear of your car at an angle with a glancing blow. If he's still in front of you and you can't stop in time, you might be able to turn in the same direction he is going so again he will hit you at an angle with a glancing blow. This is why it is so critical to be aware of where other cars are around you at all times. Is there traffic coming the other way? You don't want to make an avoidance maneuver that places you in greater danger with other traffic. Remember this is all about the Laws of Physics and the transferring of energy. We want to keep the impact energy transferring to you and your loved-ones at a minimum.

THE HEAD-ON

I was going into Turn One at Summit Point at the head of a pack of cars when I blew it and trailed the brake too much and spun the car. I suddenly found myself going backwards facing the on-coming gaggle of cars at a high rate of speed. My heart jumped in my throat as I saw the first car about to climb over my right front tire and impact my helmet killing me! It was only a split-second decision but I slammed on my brakes so my car would skid in a straight line. The driver could then predict the direction of my car and made a quick turn to avoid me. If I had left my foot off the brake, my car would have zigged and zagged right back in front of him as he was making a desperate effort to clear me. The others followed suit getting around me. When the dust settled I did a "one-eighty" turn-around and continued with the race, albeit with two corners on each of what used to be round tires.

Don't kid yourself; in a head-on when that drunken driver crosses the median strip and is pointed directly at you, you *are* going to have a fight-or-flight reaction. Adrenalin is going to dump into your system and you better be wearing brown pants. Yes, you are going to look straight at those headlights coming at you at a high rate of speed. You have to realize you are having a natural fight or flight reaction – you are looking at the danger. This is your moment of warning.

You still have OPTIONS. So, don't panic.

You have to avoid this drunkard. Where is your Escape Route? You may and probably will have only split seconds to make a decision. This is why it is critical to always Widen your Vision and keep your eyes moving around as you drive so you know what is around you at all times. Is there room on your right? There might be a car in that lane. Is there room on your left? But this might put you head-on into on-coming traffic! Make sure you don't put yourself in additional harm's way in the process.

You have to decide. Even if it means nudging a car to your right onto the shoulder of the road, the key to remember is you have to avoid at-all-cost the head-on collision. That is where, once again the Laws of Physics come in: the speed and force of the drunkard's car is going to be stopped abruptly by the speed and force of your car. All that energy can and will kill you. You may have to trade paint with someone. You may have to hit another car with a glancing blow. You are looking for any daylight even if you have to create it. As soon as you look there, you will steer there. If the drunk driver hits you along the side of your car with a glancing blow as you are getting out of his way, that is still less dangerous than the head-on. Your car will probably spin around violently but that is helping to dissipate the energy, the force that both your cars have created.

Anything but the head-on.

Millions of years of evolution are at work here. You *are* going to look at the car that is coming at you head on! You *are* going to look at the kid who runs out in front of you! That is a Fight or Flight reaction - you look at the danger. Now you have to redirect your vision to where you want the car to go. And you have to *practice* doing just *that*!

Again, BIG EMPTY parking lot: Set out four or five half-full one-gallon plastic water bottles in a row to create an obstacle and drive towards them. SLOW SPEED! Ten, fifteen miles per hour! We're working on the TECHNIQUE, not the speed! Look at the bottles as you approach and just before you get to them, take your foot off the gas, look to the right and turn. Do it again going to the left. Now, approach the bottles but don't look at them; look past them. Then just before you get to them, look at them, take your foot off the gas, then look right and turn. This is closer to what happens in real situations: the kid suddenly appears, you look at them, you immediately look for your opening left or right and turn. *Always take your foot off the gas before turning!* I'll explain more about why in the chapter on understanding How Cars Work. Bring your speed up just a little; we're talking another five miles per hour, not fifty! Allow yourself to get used to this maneuver. Look past the bottles, just before you get to them, look at the bottles, take your foot off the gas pedal, look left, and turn. Again, it's the technique you must work on, not the speed; that's just a distraction. Control your speed to keep this a controlled maneuver.

This may seem trite but this exercise is not to be taken lightly. The situations we are mimicking here, just like racing, can be life or death. You are getting used to having to react quickly and correctly. You are helping to create Motor-Memory Reactions.

This is also a great opportunity to practice using your Peripheral Vision. As you approach the water bottles, pick a point past the bottles in the same way you would be Looking Ahead down the road. Just before you get to them, look at the bottles then go back to looking straight ahead, take your foot off the gas pedal, and use your Peripheral Vision to drive around the bottles to the right. Then practice the same technique going to the left. Bring your speed up, again five, not fifty. You are learning to use your Peripheral Vision to see not only the obstacle you are avoiding but also because you have learned to see more of the road with your Peripheral Vision you are able to see exactly where you have room for your Escape Route!

In addition you can use this set-up to practice your A.B.S. stop, and A.B.S. braking and turning. Get used to the grunting noise and the stuttering sensation that your pulsing A.B.S. brakes may make.

Remember, the more you practice these techniques the more you are training your body to respond by Motor Memory. It gets easier the more you do it. You'll also feel more confident being able to move your car around in sudden emergencies.

The study I mentioned in the Intro, whether race drivers had fewer accidents on the public street, found they had more accidents. But race drivers also had fewer fatalities. And that is what is key here. When the highway suddenly turns into a pinball game and cars seem to be bouncing everywhere racecar drivers are looking for daylight. They are looking for the escape route, even if it means trading paint or door handles with someone to create it.

I'm going to describe a situation that happened to me. I want you to understand clearly this in no way is a blueprint to be applied automatically to any situation you may be confronted by. Every situation is different and you have to judge it accordingly.

I was driving a friend's rental car one evening on Lincoln, a typical four-lane boulevard in Santa Monica lined with an assortment of commercial buildings. I was in the left lane of the two lanes going north with a car in front of me and a car along side. As we approached an intersection with no traffic light, a pedestrian stepped into the crosswalk on the opposite side of the street. In California, pedestrians have right-of-way. The southbound traffic stopped. The guy in front of me must have been distracted, not paying attention, because when he looked up he panicked and slammed on his brakes well short of the intersection. I had already lifted but was not expecting the driver to short-brake the intersection so far from the crosswalk. I slammed on my brakes and surprise-surprise! The rental car had no A.B.S.; the front tires locked and with skidding screeching tires I was headed for his rear-end. The guy next to me slammed on his brakes giving me no room to my right! The only *Escape Route* was in the southbound lanes to my left. Those lanes were empty because the pedestrian had stopped the southbound traffic. I released my brake and, since all the weight of the car was on the front tires, made a quick turn to my left into the southbound lanes and into the particular lane the pedestrian was *not* standing in. I immediately applied a threshold brake bringing the car to a smooth stop somewhere alongside the rear corner panel of the car that had been in front of me. Since I was not in the pedestrian's lane, I was well clear of him. And fortunately, he did not have a heart attack. Had there been a truck or other heavy vehicle behind me in the northbound lane, they would probably have ended up in my back seat had I still been there.

When the dust settled I waved the pedestrian across. As he crossed in front of the car that had been in front of me, I continued around both of them, back into my northbound lane and on down the road. Heart rates a little higher but nobody worse for the experience.

Once you get the techniques of racecar driving down you'll see that driving is a lot like dancing. You learn to glide across the floor deftly avoiding the other dancers. And in today's world of crowded highways and distracted "panicked" drivers, you have to be very light on your feet.

WHAT'S AROUND THE CORNER?

When making a turn, more often than not, people have a tendency to allow their vision to focus too closely in front of the vehicle. This causes them to only look at the entrance or half way into the turn. They turn the wheel only enough to get to that point.

Fig. 8A

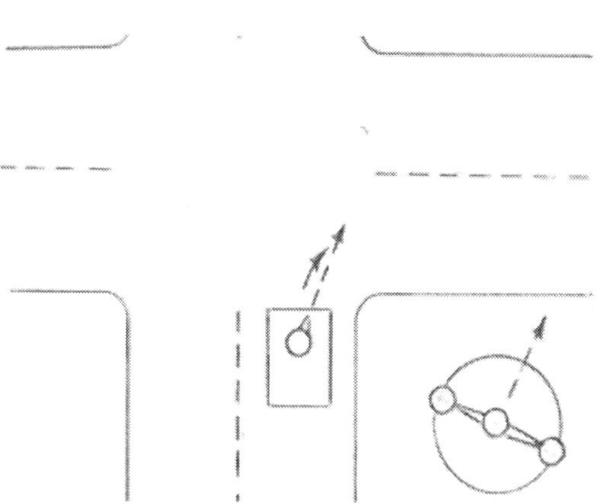

They then have to turn the wheel even sharper to finish getting through the turn.

Fig. 8B

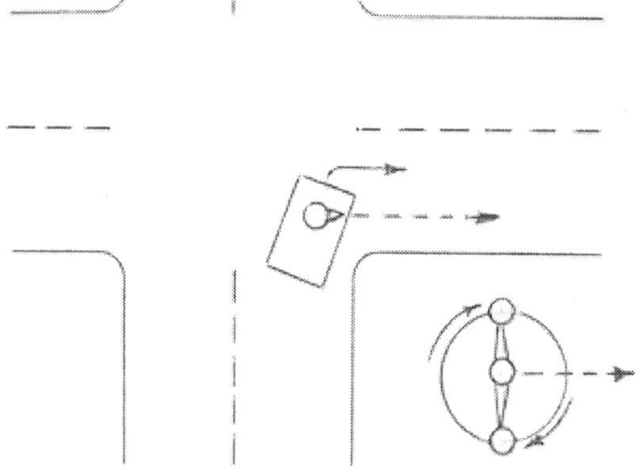

This becomes more apparent when you watch people left-turn into a double lane. They invariably end up in the outside lane.

Fig. 9A

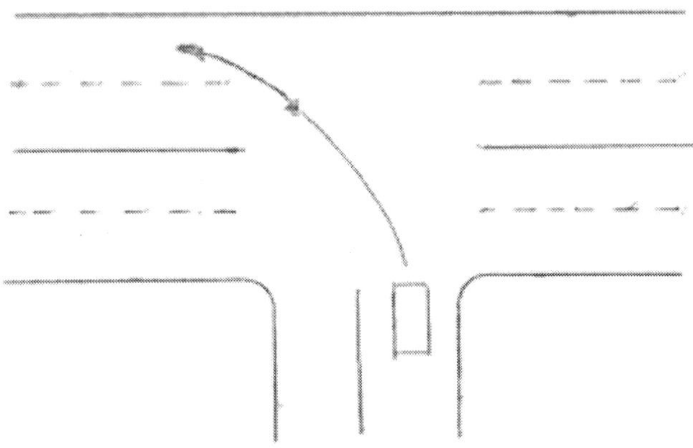

It's only when there are two left-turn lanes into a double lane the car on the inside lane will be forced by a car on the outside to turn sharper to get into the lane closest to them.

Fig. 9B

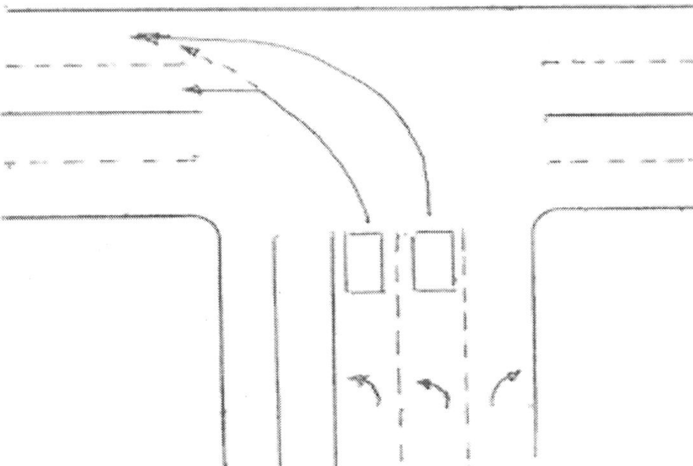

This sudden sharper turn is now asking more of the front tires and, if there are marginal road conditions like rain, it may cause the front tires to skid, "pushing" the car into the outside lane. How many times have you been on the outside of two left-turning lanes and been cut off by the car on the inside drifting over into your lane?

This is another example of why Looking Ahead and Anticipating must be applied to CORNERING. You're not only seeing if there is danger ahead but since where you look is where you steer, you are also telling your hands how much to turn the wheel. That's why it is necessary that as you ENTER a CORNER,

SECRET # 7

LOOK **ALL THE WAY** THROUGH THE CORNER.

Figure 10 illustrates Looking All the Way Through the Corner as you turn. You will complete the turn in one smooth continuous motion without putting undo stress on the tires.

Fig. 10

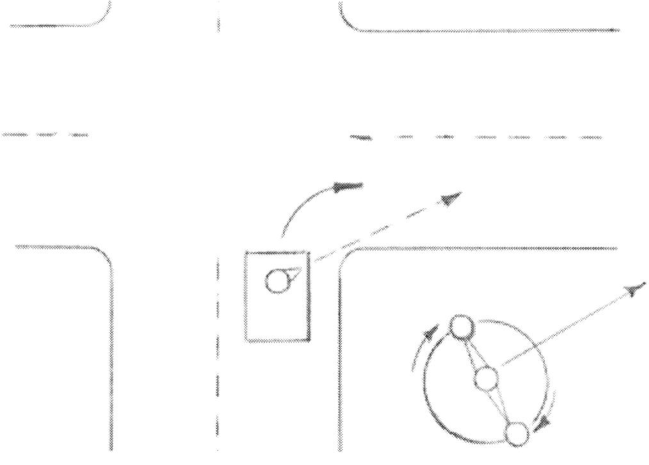

When talking about racing I've mentioned the apex point in the curve or corner. This is the inside of the corner where the race driver will come closest to clipping the curb. As he turns into a corner he is looking at the apex. Since where he looks is where he steers, he will automatically steer towards the apex. Once the steering wheel is set, he is free to look down the track to the ensuing straight away.

In other words: Look All the Way Through the Corner.

So where does the Apex come into play driving on the public streets? It differs according to the situation. On a country road where there is a dirt shoulder, you can apex the turn closer to the inside edge of the road. HOWEVER, if there is a drop off at the edge of the road, do NOT push your luck by trying to force an apex! The drop off can hook your tire and cause you to spin! You are looking for rhythm, not tenths of a second.

You want to do what is called a "late apex." As you look into the corner visualize the point where you will get closest to the inside edge of the curve *beyond* the halfway point of the corner. The solid line indicates the "late apex" in the figure below:

Fig. 11

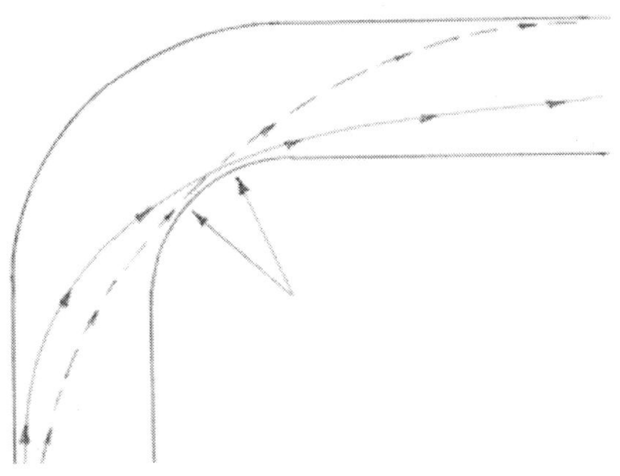

You are getting most of your turning done before you get to the inner portion of the curve. This allows you to straighten the wheel sooner, and get back to the gas pedal earlier for a smoother acceleration out of the corner.

I'll talk more about this in Secret #8 - Slow in Fast out.

This "late apex" is also very important for making a correct left-hand turn into a double lane. You are getting most of your turning done before you get to the apex as in the figure below.

Fig.12

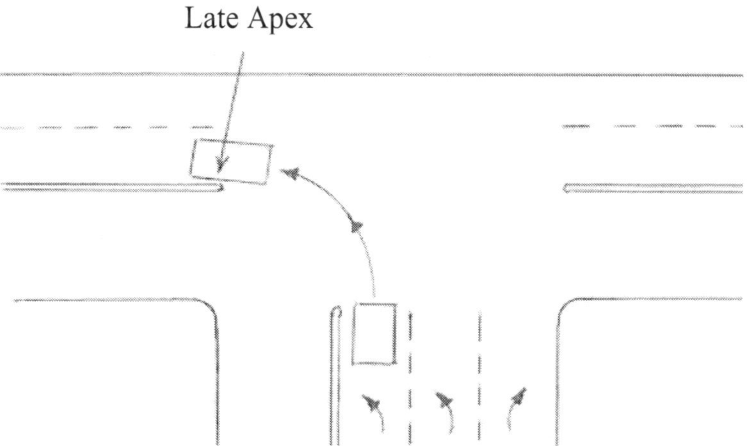

This prevents you from drifting into the outer lane and a potential fender-bender. Once you turn the wheel, don't keep looking at the apex. You will at first. But once you gain confidence, you'll begin Looking sooner All the Way Through the Corner.

In city or suburban driving, don't even think of looking for an apex where there are pedestrians. Since the passing of pedestrian laws that give pedestrians right-of-way, be leery of any pedestrian standing on a corner. More often than not, they are staring at their smart phone and not paying attention. They could step in front of you at any moment. The phones might be smart but the users, all too often, act like they are not. Give yourself at least a car-width berth around any corner where there are pedestrians.

Looking all the way through a turn applies to U-turns as well. After checking for traffic in both directions, as you are about to turn the wheel, look out the *side window* in the direction you want to go, not the front windshield. This is telling your hands how much turn of the wheel is necessary to complete the U-turn. By looking all the way ahead through the turn, whether it is a right, left, or U-turn, you are always telling your hands how much to turn the wheel. Your Peripheral Vision will help you keep track of the edge of the road or an apex if you are going around a median strip.

If it is a blind corner blocked by a hillside or in the city blocked by a building, you must certainly keep Looking Ahead as far as possible through the corner. There is a myriad of possibilities: other vehicles, bicyclists, pedestrians, and anybody in a hurry running a red light. This is where utilizing your Peripheral Vision becomes very important. Rather than trying to spot each potential threat, by Widening your Vision you are able to see any unusual movement and if that movement is beginning to cross your path. This will help free you to Look All the Way Through the Corner even in the busiest of city streets.

Look Ahead in *ALL* directions, and Anticipate to create that Cushion of Safety needed in every day driving.

Also, due to the number of bad drivers on the road:

DO NOT TAKE GREEN LIGHTS FOR GRANTED!

I don't know about you but it seems to me more and more people are running red lights. I used to think it was just Miami. Yeah Miami, don't give me the who-me routine; you know it's true. But now I see it all across the country. Especially since the advent of TEXTING! Look *both ways* before you step on the gas. Idiots do not take responsibility for their actions. They think somebody or something else is responsible. "The light was still yellow!" Or: "Suzie texted me if she should break up with Bobby…"

I'd like to add a point Midget race-driver Dave Carman once raised. Be aware we all have an optical-nerve blind spot in our vision. At an intersection, don't just glance from side to side looking for someone running a red light. It's rare but it *is* possible for a car to fall into that optical-nerve blind spot. Put your head on a swivel, as Dave likes to say, and LOOK in both directions.

Note too, when stopped at an intersection, check your rear view mirror for approaching vehicles. If the driver is looking down texting, blow your horn to get their attention to stop in time.

I know states are passing laws against texting but that doesn't seem to be stopping the number of cars I see weaving down the road only to find the driver looking down at their lap texting and not looking at the road ahead!

HOW CARS WORK

I explained that in racing, the idea is to get the gas pedal down to the floorboard as early as possible coming out of a corner onto a straightaway. Well, in order to do that I need to slow the car down enough before the corner so I can transition from brake pedal to gas pedal by the apex, the middle of the turn, or in some cases transition to the gas pedal just before the apex of the turn.

One of the critical mistakes race students make is carrying too much speed *into* a corner. By the time they get the car slowed down enough to get back to the gas pedal they are already well past the apex of the turn and out onto the straightaway. That kills their lap time and they'll probably get passed on the ensuing straightaway. That is of course if they haven't run off the track. This brings us to:

SECRET # 8

SLOW IN, FAST OUT

Carry your speed coming out of a corner not going into it.

So, what does that have to do with our everyday driving? Actually, a lot. Racing is nothing more than very efficient driving. Taken in this context, SLOWING the car before you get into a turn enough that you can EASE onto the gas pedal as you are coming out of the turn makes for a more efficient turn. That *easing on* of the gas pedal means you're saving gas as opposed to mashing on the gas pedal to get back to the speed limit. Also *easing on* the gas pedal as you exit a corner helps to settle the suspension for more grip. In order to understand what that means, you need to understand:

HOW CARS WORK.

Basically your car is a sea-saw or teeter-totter. Whether you're teetering or tottering, the car's front end or backend, whether up or down, is controlled by the gas pedal and the brake pedal. As illustrated in the figure below when you step on the gas pedal the front end of the car wants to go up and the back end of the car wants to go down.

Fig.13A

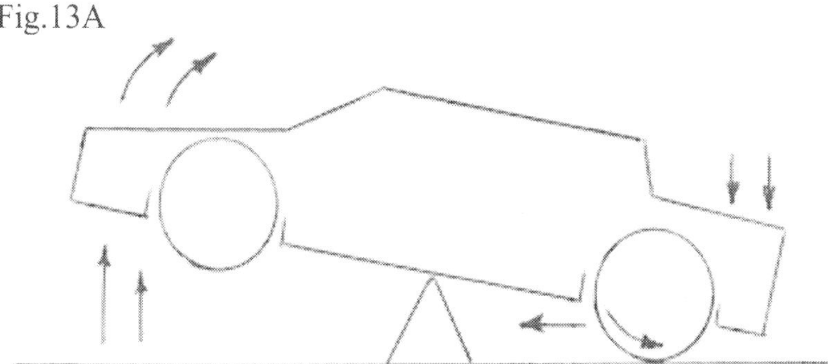

When you step on the brake pedal the front end of the car wants to go down and the back end of the car wants to go up as illustrated in Figure 13B below.

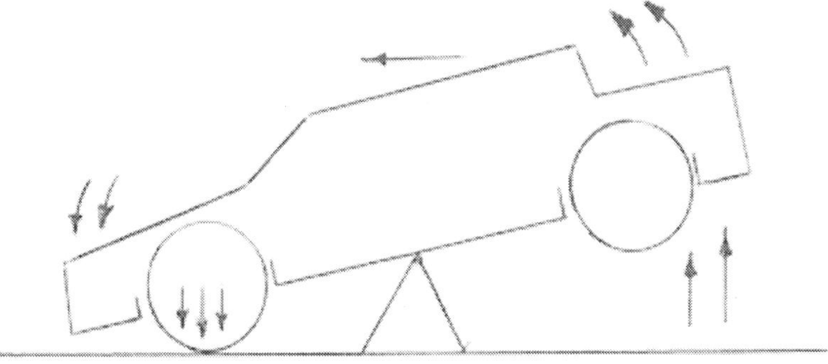

At an intersection, you can tell whether an approaching vehicle is slowing by watching to see if the nose of the vehicle is dropping. If it isn't, LOOK OUT! He might be distracted texting and is about to blow through a red light! Nose up the vehicle is accelerating. Nose down the vehicle is braking. It's a subtle difference; but the more you pay attention the better you'll get at recognizing whether an approaching car is slowing or not slowing.

To maintain maximum control of *your* car, you want to keep the car as close to the balancing point of the teeter-totter as possible. Previously, I was talking about body roll and minimizing body roll to keep all four tires evenly weighted for more traction and control. The same is true when you shift the weight (load) of the car from the front to the rear and vise versa. You want to minimize that shift, which is done by *easing* or gradually pressing the pedals *not mashing* the pedals. Mash on the gas pedal and the front end will come up taking weight off the front tires where you could lose your steering - Under Steer. Mash on the brake and if the nose of the car dives too much causing the rear end to come up abruptly, you could lose the rear traction causing the car to spin - Over Steer.

At the race schools you'll hear instructors telling you to be smooth. They're not talking about sauntering out to the car with your patented walk, tossing a scarf over your shoulder and flipping a cigarette off in a devil-may-care-give-a-rat's-patooty attitude. They are talking about being smooth with the controls: the pedals and the steering wheel. Turn the steering wheel in one smooth motion; don't jerk the steering wheel. And with the pedals: squeeze the pedals down, don't kick them down. Unless it's an A.B.S. panic stop, in which case grenade that brake pedal, use your whole leg for leverage and slam it down to the floorboard to trigger the A.B.S.

In the chapter after the next I'll talk more about the proper seating, foot and hand positions that enable you to squeeze the pedals correctly and more about the difference when it comes to A.B.S. For now, think of operating the pedals as the difference between squeezing the trigger and pulling the trigger. In the same way pulling a trigger makes the gun jump in your hand and you miss the target, when you mash on the pedals you upset the balance of the car destabilizing it and causing it to miss the road. Here comes that tree again!

This is particularly true for Porsche 911 drivers or other rear-engine cars such as early VW Beetles and the exotics like Ferrari and Lamborghini. All that weight in the rear can have a pendulum effect that can cause the car to go into sudden Over Steer, spinning the car. By easing into the gas pedal - and I emphasize EASING - it shifts weight to the rear tires gradually, increasing rear traction to handle the added weight of the rear engine without taking undo weight off the front wheels and decreasing your turning capability. All that techno talk means slow the heck down before the turn, or your 911 is going to turn into a whirling dervish that will not end well.

To a lesser degree the same can be said about any car. As you are coming out of a corner you still have the wheels turned. As I said before, if you mash on the gas and shift the weight of the car to the rear, just like a teeter-totter, the front tires are going to go up losing much-needed weight that gives them the grip necessary for the car to turn. Again, this is called Under Steer or Push. So, ease into the throttle coming out of a turn.

Going *into* a turn presents a similar problem. The infamous Corkscrew at Mazda Raceway Laguna Seca outside Monterey, California is infamous for a good reason. Aside from the fact people have died there, when you launch off the top of the hill to enter the Corkscrew, it's a three hundred foot drop from the top of the hill to the bottom. It's like diving into a thirty-storied elevator shaft with a sharp right-hand turn that transitions to a sharp left-hand turn as you hurdle downward; thus the corkscrew.

As you drive off the top of the hill, you can't see the racetrack. Many a first-timer has had a religious epiphany going off the top. The word "Holy" comes to mind usually followed by an expletive. The racetrack will literally disappear from view under the car. In order to be aimed in the right direction as you go off the hill, you have to look at the opposite hill where there are three trees. You aim at the tree on the right so when you come back to earth the car will be at the apex of the downhill right-hander. As soon as you hit the ground you make a hard right-hand turn. From there you drop into a pocket where you up-shift. In the sport sedans we use at the race school you stay on the gas pedal the whole time until you transition to the downhill left-hander. At that point you have to get *off the gas pedal* to get the *nose down* so you can get *weight* on to the *front tires* to turn left. It's complicated by the fact the track is still falling out from under the car the whole time as it continues in a steep downhill. So if you're carrying *a lot* of speed with a high horsepower racecar, you may also have to apply the brake to get the nose to drop enough so the car will turn. Once the car has completed the turn you drop into another pocket. From there you want to get the racecar pointed straight before the apex of the downhill left-hand turn or the car will Under Steer as you leave the pocket and you'll fly off the track. Someone I knew died there. Once the racecar is straight, you are right back into easing the gas pedal on full before braking for the bottom of the hill.

This may sound like an extreme case, but anyone who has driven on an expressway or highway has been in a similar situation albeit at a somewhat slower speed. See if this sounds familiar: you are leaving the highway onto a downhill circular exit that ends in a stop or a merge into a street that is passing under the highway you are now leaving. You ease off the gas pedal to slow down from your highway speeds. As you turn onto the exit you find the car "pushing" out to the Armco barrier that lines the outside edge of the road! What is the fix that will keep you from hitting the barrier? Get your foot off the gas pedal if it's still there. This will get the nose of the car to drop and put more weight on the front tires so they can turn the car. Since the road is dropping out from under the car as you go into the downhill turn, it might require a little brake to add more weight. Remember the brake shifts more weight to the front. Squeeze the brake pedal down in one smooth but quick motion. Don't over do it. If you hit the brake pedal too abruptly or too hard while you're still turning you could be asking too much of those front tires and they'll lose traction and skid. By the way, where are you looking while all this is happening? Look through the turn, maybe as far ahead as the stop or just before the start of the merge. Look as far ahead as you are comfortable and still able to see the Armco barrier with your Peripheral Vision. Your Peripheral Vision is going to tell you whether the car is headed toward the Armco or not. Besides, what happens when we look at the Armco barrier? That's right, you start turning toward it!

In extreme cases where you come off the freeway too hot and are carrying far too much speed to get the car to turn, there is still a way to save your self. A.B.S. YOUR BRAKES! Remember, it Allows you to Brake and Steer. Again, push the pedal all the way to the floorboard to activate A.B.S.

In racing there is a trick drivers use if they over-cook a corner. They have only split seconds but if there is enough room, they will straighten the wheel and quickly squeeze the brake pedal hard to slow the vehicle. At that point, they're asking the front tires to do only one thing at a time: just brake, not turn *and* brake since they've already passed the adhesion limits of the tires and gone to massive Under Steer. This braking in a straight line helps put more weight (load) on the front tires to get the vehicle to turn. Then, as they are easing off the brake pedal they again turn the steering wheel back into the corner allowing the racecar to continue. This is a demonstration, albeit an extreme one, of Race Driving's Secret #9 that shows you: *How to get your Feet to follow your Hands*.

SECRET #9

I work with an automotive marketing company showcasing vehicles to the public. It's great fun because I get to drive all kinds of cars and then explain the advantages of the vehicles as members of the public drive them. I also get to see how a lot of different people drive. Sitting in the right seat I've noticed a very curious thing. As people are entering a corner, they turn the steering wheel and STEP ON THE GAS PEDAL at the same time. This can cause the car to Under Steer or Push. When I ask about it, they have no idea they are doing it. What they are doing is actually a mild form of fight or flight, a nervous reaction, and it is the opposite of what they need to do to keep proper control of the car.

Your *foot* movement must follow your *hand* movement properly. We already know that where we look is where we steer; our hands automatically follow our eyes. We look where we want the car to go, our hands automatically steer there. The next thing we have to do is get our feet to follow our hands. This however is not automatic. Given how quickly the fight or flight reaction can come into play even in the most benign of circumstances I would have to say it's just the opposite - we're more inclined to do the wrong reaction. To get your feet to follow your hands correctly requires thinking and practice. And that brings us to a Race Car Driver's:

SECRET # 9

EVERY TIME YOU TURN THE STEERING WHEEL, YOUR FOOT COMES UP.
EVERY TIME YOU STRAIGHTEN THE WHEEL, YOUR FOOT GOES DOWN.

Here's how it works: Imagine you have a string tied around your foot. At the race school we would tell the students we were going to sink an I-bolt through their foot and attach a string to it. This usually elicited uneasy laughter. We are now going to take the other end of the string and attach it to the bottom of the steering wheel. Place your hands at nine and three o'clock on the steering wheel. As you turn the steering wheel, it pulls the string upward lifting your foot as in Fig. 14A. When you straighten the steering wheel, it lowers your foot as in Fig. 14B. Every time you turn the wheel, your foot lifts up off the pedals. Every time you straighten the steering wheel, your foot is free to come down on the pedals.

Fig. 14 A

Fig. 14B

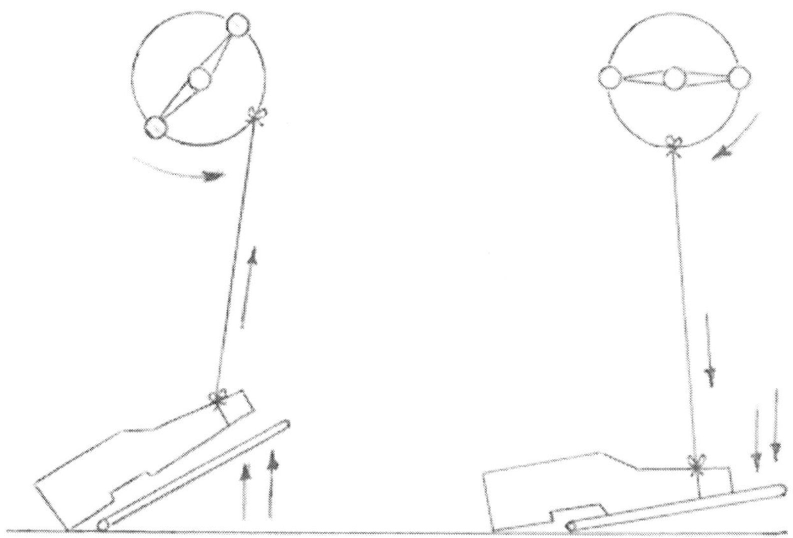

You are driving down a road and you are coming to a corner; your foot is still on the gas pedal. You lift it and squeeze down on the brake pedal to slow the vehicle for the corner. As you are about to turn into the corner, your foot *eases* off the brake pedal. You don't want to take your foot off the brake pedal abruptly. It may cause the front end to bounce up – like a teeter-totter – taking the weight off the front tires resulting in possible Under Steer. Make sure your foot is completely off the brake pedal as you begin turning the wheel. Think of it as putting imaginary slack in the imaginary string that will allow you to turn the steering wheel. As you are turning the corner, just past the midpoint of the turn, the apex, shift your foot over to just above the gas pedal. As you are coming out of the corner and gradually straightening the steering wheel, think again you are putting imaginary slack in the string. This allows your foot to come down slowly on the gas pedal accelerating smoothly and gradually out of the corner. Once you do this over and over, motor memory will begin to take over and it will become automatic.

And what's happening to those tires all this time?

The average tire patch "where the rubber meets the road" is about the size of a man's hand with the fingers stretched out. All four tires give you a total tire patch of only about 1.3 square feet.#9 That is what is controlling your three to five thousand pound vehicle! So, don't be asking that tire patch to give you full brake or full acceleration *with* full steering or it will lose traction and skid. If you are braking 100%, you can't apply steering until you've put slack in that string by getting the braking down to 10% or less.

I AM NOT SAYING DON'T TURN AND BRAKE AT THE SAME TIME!

Average posted street speeds are well within the adhesion parameters – grip – of most tires even when braking and turning at the same time. What is most important is getting the car slowed enough so the car *will* turn.

If you have 100% steering coming out of a corner, then don't apply 10% accelerator until you've gotten to 90% steering. As you are gradually straightening the steering wheel, you are decreasing the percentage of steering – 50%-40%-30%-20%-10%-0%. *That* allows the gradual increase in percentage of accelerator pedal - 10%-20%-30% - UNTIL THE SPEED LIMIT (DAW)!

You can ask the tires to fully do only one thing at a time. But by alternating degrees, you can ask your tires to do two things at a time if you gradually decrease one as you increase the other. Again, what happens when I ask them to do too much of two things at once? It's possible they can skid.

Race driving has the tires constantly at the their adhesion limits. So to turn, a race driver *must* slow the racecar enough to where he can gradually give up the braking demand on the tires as he gradually increases the turning demand on the tires. As he exits a corner, he is easing as quickly as possible into the accelerating demand on the tires as he gradually unwinds the turning demand on the tires. Like I've said, race driving is just very efficient driving. So why not apply that efficiency to everyday driving?

Again, when coming to a corner, brake in a straight line; *ease* your foot up off the brake pedal; then turn in looking through the corner. When coming out of the corner, *ease* the gas pedal back on as you are straightening the wheel. You want to be as smooth (think gentle but firm) with the controls as possible. This keeps the weight of the car balanced from front to rear giving you more grip and *more control* for a greater margin of safety.

PROPER POSITIONING

Now in order to be smooth with the controls you have to start with proper seating positioning. I know there's a lot of guys out there who think being smooth is having the seat all the way back with their arm stretched out in front of them and their fingers draped over the top of the steering wheel. Unless you are cruising at ten miles per hour, you are a collision about to happen. And even at ten miles per hour, there's no guarantee.

Proper everyday seating position is the same as proper race seating position and it starts with the legs. There should be bend in the knees. In fact, I want you to put your foot on the floorboard behind the brake pedal and check to make sure there is still a slight bend in the knee. If there isn't and your leg is straight, when you slam on the brakes and impact in a collision, your knee will snap backwards breaking your leg. So move the seat forward enough to get an appropriate bend in the knee. This bend in the knee is also important to avoid unintended acceleration. I'll talk more about that later. Now that the seat is positioned, make sure it is raised high enough so you can see cleanly over the dashboard. Nothing drives me crazier than to see petite women or the elderly staring down their nose as they strain to see over the dashboard. RAISE your seat! If this means you can't reach the pedals then move the seat further forward.

Next, there should be bend in the elbows as you grab the wheel at the 9 and 3 o'clock positions. Yes, 9 and 3. Not 10 and 2. Your hands anywhere above 9 and 3 and you will know exactly what time the collision was because your wristwatch will be embedded in your face! When the airbag goes off it will blow your hands right back at your forehead. At 9 and 3 your hands will get blown back toward your shoulders.

There is also a physiological reason for 9 and 3. The higher up you grip the wheel, the more your shoulder muscles will tighten.

You'll realize this on a long trip when your Trapezius muscles between your shoulders and neck start to cramp. The lower you grip the steering wheel the more your arms and Lats, the muscles at the outer edge of your back below the shoulder blades, will tighten. With your hands at 9 and 3, you are using all your upper body muscles in unison: chest, shoulder, back, stomach, arms both triceps as well as biceps. It's not about muscling the car; it's about allowing all your muscles to relax since you're not over-taxing any one group of muscles.

Also, 9 and 3 allows for a smoother turn of the steering wheel because you'll have more leverage on the wheel. Guys who put their hand up at 12 o'clock have a tendency to jerk the wheel when they steer in an emergency. The same is true if your hand is reaching across the steering wheel. Also do NOT put your hand up inside the wheel. I often see women doing this trying to get better leverage as in Fig.15A. Should you be involved in a collision at that moment, the steering wheel could snap back breaking your wrist. Keep your hands on the outside of the wheel at 9 and 3 as in the second figure: Fig.15B.

Fig.15A

Fig.15B

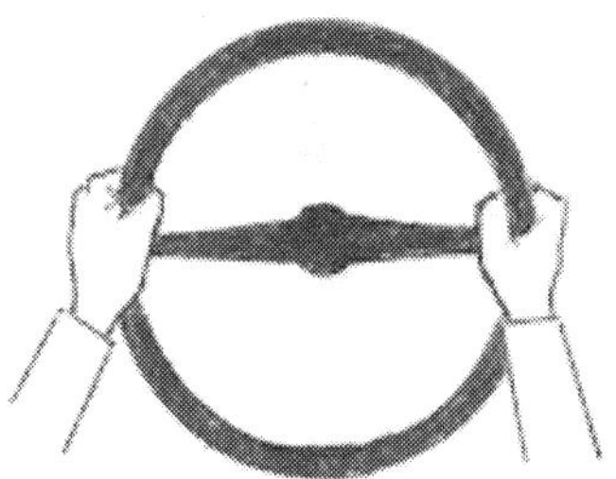

Be aware that in most vehicles, the steering wheel is canted away from you – the top of the steering wheel is farther away than the bottom of the steering wheel. There has to be enough bend in the elbows to allow the slack necessary to make a full turn on the steering wheel without pulling your back off the seat. A quick way to establish proper position is to rest the inside of your wrist on the top of the steering wheel at 12 o'clock without your back coming off the seat. This generally will give you proper bend in the elbows with hands at 9 and 3.

Which brings us back to seat position: sit squarely in the seat. The seat doesn't need to be straight up, unless you are in a dirt track Sprint car. It can be canted back slightly for comfort and I emphasize slightly. Think comfortable not couch potato. What is key here is that you are centered on the wheel with your back firmly in the seat, not leaned to one side or the other and not with your back off the seat. A lot of drivers will cant the seat back, but then sit up with their back off the seat. If you swerve suddenly in an emergency, since your back is not being held in place by the seat bolsters (the padding on the side of the seat that sticks out), your body will be thrown sideways by lateral G-force in the opposite direction of the turn. That means you are now hanging on to the steering wheel for support rather than steering the wheel for control. As you pull on the wheel to upright your self you will be turning back into the problem!

Race drivers are so firmly strapped into their seat; the only thing they *can* move is their arms and feet. Sit with your back pressed comfortably but firmly against the seat, hands at 9 and 3 with bend in the elbows and bend in the knees and, like a race driver, you are ready for anything!

KEEP THAT HEEL ON THE FLOOR

Just like there is a proper hand position there is also a proper foot position and I am talking about your right foot, the one that should be controlling the pedals.

For all you left-foot or wannabe left-foot brakers: *unless* you grew up on go-karts or you can alternately put left foot down while bringing your right foot up or bring your left foot up while putting your right foot down without thinking about it --- fugedaboudit. You can always tell the left foot brakers going down the highway; their brake lights are on. If you are wearing out your brakes at an unusually fast rate, you might want to pay attention to where your left foot is resting.

The more serious problem for left-foot brakers is that in a panic stop they are more likely to hit both the brake and the gas pedals at the same time! This negates the brake pedal and puts you into a collision! That same mistake in a racecar has been known to kill the driver. So, keep the left foot on the dead pedal, the angled space located on the left side of your floorboard, and not the brake pedal.

Place your right foot so your heel is on the floorboard halfway between the brake and gas pedals. You might want to fudge it a little toward the brake. With your heel on the floor this enables your foot to SQUEEZE the pedals down. Remember you want to be smooooooooth! It's the difference between pulling the trigger and SQUEEZING the trigger. Use your thigh muscle to keep your heel positioned on the floor. You are now using your calf muscle to operate the pedals. That allows for a smoother application of the brake and gas pedal and makes it less likely the car will pitch forward or back. Remember the car is a teeter-totter. The only time you use your whole leg, your thigh as well as calf muscles, to operate the pedals is when you A.B.S. the brakes in a panic stop!

I want to point out here that not everyone's pedals are going to be the same. Yeah, another uh-duh moment. Most brake pedals are mounted on the firewall so they hang down. With this type of pedal mounting it is generally easier to squeeze the brake and gas pedals while keeping your heel on the floor. With pedals that are mounted in the floorboard, it becomes more difficult. It may require you to drag your heel forward as you apply brake or gas. This motion will require using your thigh muscle. Just realize your thigh muscle can produce large movement on the pedal as in the case of a panic stop and the application of A.B.S. Your calf muscles, which are smaller, can produce smaller more graduated movement. You want to make sure you are moving the brake pedal with your calf muscles and using your thigh muscle to only position your heel, not move the pedal. For those of you with a stiff or hard-to-push brake pedal, your thigh muscle may have to come into play. Just remember it produces large movement, so be careful not to make sudden movements, but graduated movements. Of course, we are not talking the slow-drip motion of molasses in winter, here. The speed of your brake application is *as needed.* But if you can keep in mind and can constantly practice that it has to be a graduated movement of the pedals, then you are ready to learn Secret #10.

SECRET #10

USE THRESHOLD BRAKE.

This is a race driving technique that enables you to stop in the shortest possible distance. Perfect your squeeze and you can stop faster (read: shorter distance) than with most A.B.S. systems.

Practice in a parking lot, not on the street. Again, the guy behind you won't be ready for your sudden stop and may hit you! Start with a 1-8 count, #1 being your foot resting on the brake pedal and 8 being the pedal down as far as you can push it. At a standstill, count and concentrate on pushing the brake pedal down with a smooth consistent motion until you can't push any further. This is the *easing* or *squeezing* of the pedals I've been talking about.

Once you've gotten the feel for it and built a little motor memory, set up a marker like a one-gallon water bottle. Use a consistent SAFE speed for the size of the lot. Remember, it's not about the speed; it's about the technique. Start your braking when the water bottle is next to your door. When you've stopped, set out another water bottle next to your door. Now, try it again at the same speed using a 1-5 count with 5 being the pedal pushed as far down as you can. Compare the distances. Try it again with a 1-3 count, without skidding or triggering the A.B.S., and your braking distance will be even shorter.

Note: if you start skidding, don't jump out of the pedal, ease out of the pedal until the skidding stops then squeeze down again and come to a stop. The idea is not to ask so much of the tires they loose traction and skid. The stop should be smooth and quick without skidding. Once you get your smooth application of the brake down, it will become motor memory and you'll find yourself SQUEEZING a Threshold Brake every time, just like a race driver.

BEING SMOOTH APPLIES TO TURNING THE STEERING WHEEL

Race driving as I've said is very efficient driving. One such efficiency in driving is conserving energy (read: save gas) by saving momentum. And that brings us to:

SECRET # 11

WHEN YOU TURN the WHEEL,

YOU **SLOW** the CAR DOWN.

Compare the larger and smaller arcs below and the amount of steering wheel input it takes and the comparative speeds:

Fig. 16A Fig. 16B

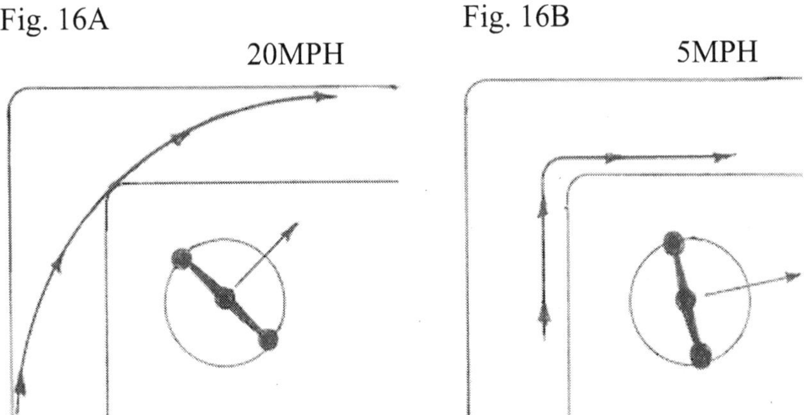

In racing we minimize the turning of the steering wheel to maintain speed or momentum. We use just enough steering input to get the job done. Remember Watkins Glen Race Track and the turn at the end of the back straightaway? I started from the outside edge of the track, turned down into the inside edge at the apex and used the entire road as the car slid out to the outside edge on exit to take the turn flat out. By making the largest possible arc through a turn you minimize the steering wheel input to maintain speed and save momentum.

In your everyday driving the entire amount of road available is just your lane. If you are on a country road the lane may be wider but in the city it will definitely be limited. By creating the largest arc through the turn while staying safely in your lane, you can accomplish a turn that minimizes steering wheel input and saves momentum. You're also minimizing wear and tear on your tires and saving gas at the same time. Stop and think about it. There's a reason you have to replace your tires. Every time you take your car out for a drive, you are leaving a little bit of your tires out there on the road surface. Unless you lock up your brakes for a screeching stop or do a "burnout" on acceleration, you're not going to see it, but it's there. This is particularly true of your front tires. They wear out faster because they are constantly scrubbing off more tire rubber as you change the direction of your car when turning. That's why many car manufacturers recommend rotating your tires – moving your more worn front tires to the rear and putting the less-worn rear tires on the front. There are depth gauges to measure tread wear on your tires to know exactly when to change them.

Turn the wheel enough to get the job done safely without over doing it. Add to this a shuffling motion starting with your hands at nine and three o'clock as in the figure below.

Fig. 17

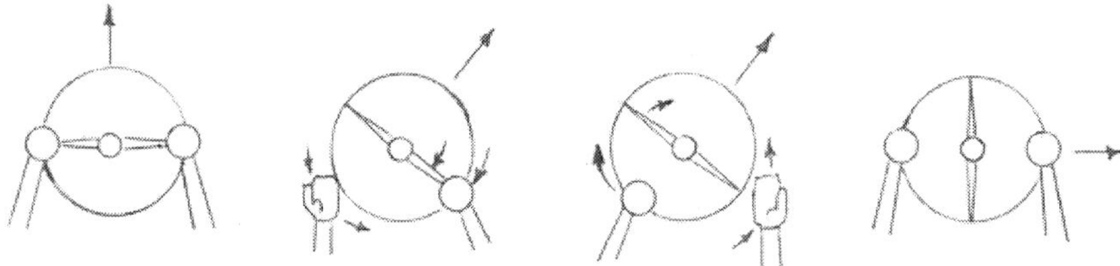

Shuffling keeps your hands in the proper nine and three position, which gives you more leverage on the wheel. Try it in a parking lot; you'll be surprised how quickly and smoothly you will be able to turn. Look Ahead through the turn to tell your hands how much to turn the wheel, Anticipate, Breath, Keep your shoulders, arms and hands Relaxed - no death grip on the steering wheel. And by shuffling you won't be over turning the steering wheel creating the unnecessary resistance that wears out your tires and burns more gas!

Note: there is a current school of thought to keep your hands at nine and three on the wheel and to cross your arms over in a turn until you can't go any further and then do hand over hand to complete the turn.

My personal preference is shuffling the steering wheel because it feels more comfortable, makes for a smoother turning motion, and makes it less likely to turn the wheel too much. But try both methods and see which feels the best to you and gives you the smooooothest turn of the wheel.

An additional word about SAVING GAS: keep your engine speed below 2,000 RPM. You can read this on your Tachometer, the gauge generally next to your Speedometer. If you shift below 2,000 RPM (engine *revolutions per minute*), you are using more of the transmission to accelerate the vehicle rather than the engine. This saves gas. In racing it is known as "short shifting" and is used to save gas when close to the end of the race.

UNINTENDED ACCELERATION

Why it happens and what to do when it happens to you.

Nobody likes a surprise unless it's HAPPY BIRTHDAY! And yet everyday, probably thousands of times a day, all across the country, people get surprised behind the wheel by Unintended Acceleration. Most cases produce minor fender benders. The ones that hit the local TV news or YouTube are the ones that go through a storefront. But the ones that plow through a crowd of people hit the national news. We've all heard of the cases of the little old man or woman who plowed into a storefront or ran amuck through a crowd of people. Believe me it is not just the elderly who are susceptible. It can happen to any one, any age, anywhere. No one is exempt. And all of it is for the same reason.

I was teaching an emergency driving school, and at the lunch break one of the students, a middle-aged man, asked me about a situation that happened to him. He was driving an older luxury car – one without Electronic Stability Control. He was on the highway when the guy in front of him slammed on his brakes. He swerved to the left to avoid him. As he tried to straighten the car, it went into massive Over Steer! He corrected by turning into the skid and the car snapped back in the opposite direction again in massive Over Steer! He corrected again, and again it snapped back in the opposite direction in massive Over Steer, often referred to as "Tank Slapping!" This went on and on, Over Steering in one direction and then swinging back wildly into Over Steer in the other direction until finally he gave up. He said he put it in God's hands and took his hands off the wheel! Apparently God was looking out for him. He shot across three lanes of traffic, hit a retaining wall, totaled the car, and walked away from it! He then asked me quite seriously, "What did I do wrong?"

I asked, "You mean, besides taking your hands off the wheel?"

He rolled his eyes and nodded.

I answered him with, "You didn't take your foot off the gas pedal."

He thought about it, reliving the event in his mind and finally answered, "Oh my god, you're right!"

I know there are tragic cases of people, family members, friends who have died because of Unintended Acceleration. I don't want what I am about to say to be misconstrued or misunderstood as trivializing in any way the horrible suffering that families and friends have endured with those confronted with Unintended Acceleration. For those victims who survived, they swear their foot was on the brake. For those who did not survive, the grieving families are convinced something was wrong with the car. The families are looking for some kind of explanation, and they want someone to pay, some car manufacturer to pay for these tragic cases. Unfortunately, after extensive investigations, the cases almost always come back to "driver error." How can this be? Well, the fact many rear-end collisions show no sign of skid marks to indicate braking may be the clue.

Here is what is happening: remember our common Ancestor? The one who handed down the fight or flight reaction? The one who looked at the danger and then JUMPED?

Well, it comes down to this: Unintended Acceleration is almost always the result of the JUMP REACTION in the FLIGHT portion of "fight or flight."

Anything can set off the flight-and-jump reaction behind the wheel: you're not paying attention, you're twiddling with the cell phone or radio, you look up to see the car in front of you has suddenly stopped. And you jump! You are driving on the highway at highway speeds; suddenly a car in the next lane swerves, and you jump! That sudden jolt of fear in your body - your arms stiffen, your hands tighten on the steering wheel, and your feet jump! It can be any sudden surprise that you think threatens you and you will look at the danger and your feet will jump – the flight reaction - just like when lightening strikes or somebody jumps out at you and yells BOO!

The problem is when you jump; your feet don't jump inward, THEY JUMP OUT! The better to launch you to run from danger! But you're in a car. When your feet jump out, guess what pedal you are going to hit? Which pedal is on the outside? The gas pedal. You'll hit it every time. Remember, this is instinctive; you have no control over your feet. Your foot has jumped on the gas pedal, but here's what makes the problem worse: in your mind, you think your foot is on the brake! The universe is suddenly upside down and you don't understand why the car isn't slowing down so you press harder. You are in full-on-get-me-away-from-the-saber-tooth-tiger-flight mode! You are in full-on panic!

I had a guy I was coaching around a racetrack during a Corporate Bonding Program. The final exercise was a competition among the participants. It was a simple set up: two opposing teams, two mirrored tracks. Each participant raced around the track, came to a stop, jumped out, and tagged the next participant who jumped in and started the process all over again. The first team to get all their drivers around the track won. I was right-seating and coaching them as they went around our track: where to be looking, when to brake, and when to accelerate.

This particular young man was the last one to compete. By this time, with all the ongoing cheering and catcalling he had pretty much worked himself into a frenzied state. I clicked his seatbelt, the instructor outside closed his door, I told him to put his foot on the brake, I dropped us in gear, and off we went.

He was doing fine but I could see he was starting to tighten up as his eyes got bigger and bigger. I pointed the way and told him to squeeze on the throttle. As we came up to the last corner before the finish line, I told him to get ready to brake, and told him, brake, brake, and yelled, BRAKE!! I hit the gear shift into neutral, yanked the E-brake, reached across and yanked the steering wheel down to spin us out and keep us from going through a tent full of people. When we spun to a stop in a cloud of dust, the engine was screaming at full speed and bouncing off the rev limiter. When the dust settled, I told him he could take his foot off the gas pedal now.

He looked at me dazed and said, "What?" He looked down and saw his foot was on the accelerator and yelling, "Oh my god!" took his foot off the gas pedal. He looked at me and said, "I thought my foot was on the brake!"

I smiled and said, "…I know."

Here's how the human body works in fight or flight: physiologically you are loaded with adrenalin and running like heck from the saber tooth tiger - you are operating off the millions-of-years-old-portion of your brain. Without getting into a protracted explanation about the amygdala, the hypothalamus, and the inner workings of the brain, I'm going to simplify the explanation to just "right-brain" and "left-brain." The right brain is the portion of the brain that can produce the flight reaction and the potential for panic. The left-brain produces logic. Triggering left-brain activity through the use of logic can get you out of your right brain. One such logic exercise is counting backwards from five hundred by sevens – four hundred and ninety three, four hundred and eighty six, four hundred and seventy nine, and so on - you have to think logically to do this. This can take you out of the flight reaction and the potential for panic. Try it next time you feel nervous, scared or anxious about something.

I've been in enough gut check situations in racing to know that sensation of panic coming on. But I know enough about cars to know that I always had options. One such situation involved turning the steering wheel in a high-speed corner at Summit Point Raceway and getting no response from the front wheels. My heart jumped into my throat but I yanked the wheel further to get a safety bolt in the steering column to catch and the car finally turned.

There is nothing hocus-pocus about cars, nothing supernatural regardless of the *Christine* movies. A car simply does what you tell it to do. It's just a matter of telling it the right thing. That requires you getting out of your right brain, getting into your left-brain and figuring out *what the heck is going on!*

It all starts with recognizing you are having a fight or flight reaction. You have to recognize the problem before you can solve it. If you are feeling panicky, you are going into right brain. You are having a natural fight or flight reaction to a scary situation. You probably won't have time to count backwards to get out of your right brain and into your left. But you have to get yourself under control to get the car under control. So if the car does something that scares you and you don't know why, just before the panic sets in, remember this:

SECRET # 12

WHEN IN DOUBT – BOTH FEET **OUT**!

If the car starts accelerating and you don't know why, LIFT YOUR FEET! Get your feet off the pedals. UNWANTED CAR ACCELERATION plus the ONSET of PANIC means LIFT YOUR FEET! More than likely, you had a "flight" JUMP reaction and your foot is on the gas pedal. Both Feet Out, in other words Off The Pedals, and the car will settle and slow. This will allow you to get yourself together and take steps to regain control of the situation.

You might want to try a simple exercise. Big empty parking lot, slow speeds! Accelerate, and then take your feet off the pedals abruptly. This getting on and off the pedals happens all the time in our every day driving but we don't give it any thought. Well, now give it thought. Pay attention to the rapid deceleration with the deliberate lifting of your feet. Feel how quickly the car slows. Do it over and over again - motor memory. The more motor memory you can establish, the more likely you are to respond with confidence in handling a situation that may require: WHEN IN DOUBT, BOTH FEET OUT.

Getting to the SOURCE of the PROBLEM, a great way to help control the "flight" JUMP reaction in your FOOT is to practice race driving foot position. Before, I was talking about positioning your foot for Threshold Braking and Squeezing the Brake Pedal. Well, now, we're going one step further. You've got your right heel on the floorboard half way between the brake pedal and the gas pedal. Keep the ball of your foot hovering above the pedals. Don't use your ankle to position your foot over one pedal or the other. If you use your ankle, your foot moves around too loosely without proper leverage and control. You are more likely to JUMP in a fight-or-flight reaction. We want to reprogram our body motions through Motor Memory. Using your thigh muscle to keep your heel anchored on the floor, use the heel of your foot as a pivot point. Keeping your ankle straight without twisting, move your knee to the right. This will pivot your foot directly over the gas pedal. Now move your knee to the left and this will place your foot directly over the brake pedal as in the Figures below:

Fig. 18A

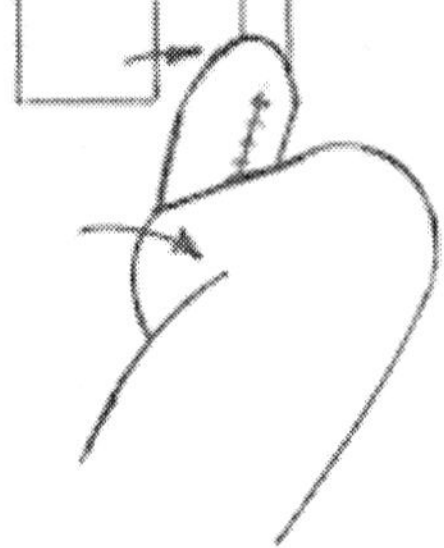

Fig. 18B

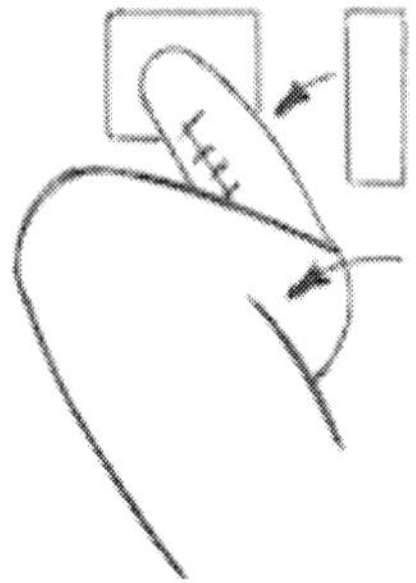

If you practice this enough, it will become motor memory. Psychologists suggest forty such movements help to imprint motor memory. You are reprogramming your body's response to a braking situation. Every time you have to brake the car, or should you need to make a panic stop, the automatic movement of your knee to the left with motor memory will insure you are on the brake pedal and not the gas pedal, thereby eliminating Unintended Acceleration.

Stability Control, by the way, which helps you regain control of your car when you skid, will actually shut off the throttle when the car skids because, more often than not, people hit the gas pedal. That flight-jump reaction! To help control it, keep that right heel on the floorboard and use your knee to position your foot over the pedals. With practice you will develop the braking technique used in racing that will help you overcome Unintended Acceleration and may very well save your life.

Okay that being said, let's say the universe does turn upside down. The car is mysteriously accelerating and you followed Secret # 12 – When in Doubt, Both Feet Out. And it's still accelerating! The first thing you have to do is get yourself under control. Remember, the car does what you tell it to do. When *YOU* ARE OUT OF CONTROL, YOUR CAR IS OUT OF CONTROL! Get yourself under control, start thinking, get back into your left-brain. You were under control enough to get your feet off the pedals. I know the next reaction normally is to step on the brake pedal. If you have a low horsepower four-cylinder engine, your brakes will probably be able to stop the vehicle. But that doesn't solve the problem. Your engine will still be trying to accelerate the car. If your car has any horsepower at all, like a V-8 or many V-6s, or you are at highway speeds, the engine will override the brake and you'll just keep accelerating. Don't panic, you have OPTIONS.

You need to take the engine out of the equation. You have two choices to do just that.

The first choice is to take the car Out of Gear. You have a transmission that is putting all that engine power to the wheels. Simply shift your transmission to NEUTRAL. If you have a Manual Transmission, simply push in the Clutch Pedal. If you are driving a Manual you will know what that is. If you have an Automatic Transmission, which today is most vehicles, simply move the Shift Lever from DRIVE to NEUTRAL. In almost all vehicles it is only one notch away. Again, practice. With your foot on the brake pedal and your foot off the gas pedal – the car should be at idle – practice moving the shifter from Drive to

Neutral. Then, when comfortable, practice in a BIG EMPTY parking lot at SLOW SPEED – 10 MPH! Your foot is on the gas pedal slow and steady, 10 MPH. You move the shifter from Drive to Neutral. You will hear the engine gain speed because it is no longer pushing against the weight of the car. The car will slow and you can then bring it to a normal stop.

Keep in mind at highway speeds should you have to do this in a real emergency, when you take it out of gear – shift to neutral – the engine will continue accelerating, get loud, and make a weird popping sound. It is bouncing off the rev-limiter - an electronic device that keeps the engine from blowing up. Don't be frightened by the noise. The engine is accelerating but the car is not. The car *is* slowing because it is no longer in gear. Just apply your brakes and get the car to the shoulder or somewhere away from the flow of traffic.

All right, the universe is really upside down! You've taken your feet off the pedals, pushed the shifter to Neutral or it seems stuck in gear and the car is still accelerating! Guess what? You still have another choice, but be aware that this is a last-ditch choice: Shut the engine off.

If you have a key that you turn in the ignition switch to start the engine, then simply turn the key back off one click. And this is why it is a last-ditch choice, be aware that if you turn it back two-clicks as when you are getting out of the car, you will lock the steering wheel and in many vehicles will shut off your Airbags. We don't want to do that with the car still moving! While sitting still and not in gear, practice turning the key back one click. Hear the engine shut off. If you have a very quiet engine – some of the newer cars have such great interior insulation you can't hear the engine – then look instead at your Tachometer. Most vehicles have them now; it's the big gauge on your dashboard that tells engine speed. It's usually located next to or near the Speedometer. If any doubts, ask your dealer or any car mechanic or for that matter any kid who likes to work on cars. The first time they put DVD players in the back seats of vans I had to have an eight year-old show me how to turn it on. The Tachometer will go to ZERO when the engine shuts off. If you should accidently take it back two clicks and the steering wheel locks just turn the key back one click and feel the steering wheel unlock. If you have one of those older vehicles where the key will sometimes jam against the locked steering wheel, you probably already know to jiggle the wheel slightly to release the key.

When you feel comfortable and you are hearing or seeing the engine shut off, take it out to a BIG EMPTY parking lot and practice shutting the engine off while you are moving. SLOW SPEED – 10 MPH! Keep your foot on the gas pedal with a steady speed; you don't need to be accelerating out of control to practice this. Again, get comfortable with shutting the engine off while the car is moving – feeling the one click and not the two. The more you practice this the more prepared you are should you need it; for instance, if your floor mat gets shoved up under the accelerator pedal and the pedal gets stuck.

Note: many OEM - original equipment manufacturer - mats have hooks in the driver side floor to keep the mats from sliding forward when you get in and out of the car. Many After-Market mats do NOT. Be aware of this if you are using After-Market mats or you have an older vehicle that is not equipped with hooks to stop the mat from moving as you get in and out of the vehicle.

If you have a push-button starter, unless it is an old sports car with a key and a push-button starter, that button is a start/stop button. You push it to start the engine and you push it to stop the engine before getting out. Because the key (fob) is in their pocket many new-to-push-button car owners forget to shut off the engine before getting out. They're in the store or at home and the car is still running at idle! You push the button to stop the engine. If you are moving, most new cars are safety-equipped to shut off the engine by simply holding the start/stop button in. Some require hitting the button twice in a row. Push the button to start the car and with your foot on the brake put it in Drive just like you normally would to go somewhere. With your foot on the brake and the car in gear, push the start/stop button twice or try holding it in. The car should shut off. Turn the steering wheel to make sure it is not locked. Some cars will shut off the power in the power steering so the wheel may feel heavy. We just want to make sure it is not locking. Then it's off to the BIG EMPTY parking lot for the next step. Again, SLOW SPEED – 10MPH. Keep your foot steady on the gas pedal and as the car is moving, push the start/stop button twice or hold it in. If the engine should not shut off - don't panic; this situation is no different than stopping at a stoplight. You are just trying to find out what your car can and cannot do. Check with your dealer to see if the car is supposed to have the safety system and if there is a problem.

The more you know about your options and the more you practice them, the less likely you are to panic. Even if you do start to panic – it's a fight or flight reaction, a red flag warning of danger – knowing you have options will help you get that panic reaction under control. Control yourself; you control your car.

Race drivers are now being taught "bilateral stimulation." One exercise is where they touch right hand to left knee, left hand to right knee, right hand to left shoulder, and left hand to right shoulder. This simple exercise repeated several times opens pathways connecting right brain and left-brain helping the drivers to perform optimally behind the wheel with more clarity, focus and thus less likely to panic in a fight or flight reaction.

THE LINE AND THE ZEN OF CAR RACING

Racing is a lot like connecting dots. You have a braking point, a turn in point, an apex and an exit or track-out point that then leads to the next braking point. The Line therefore is the connecting of those dots, the fastest easiest way around the track. Drivers sometimes refer to this as hitting their "Marks." If you go off the Line or miss your Mark you are going slower and making it harder for your racecar. That's not to say there is just one Line; like anything else in life, the Line can change. It can be affected by rain or oil on the track, tire rubber build up, or the car you are about to pass.

The Line is not marked on a racetrack. But to figure out the Line, allowing for elevation changes and the general track condition, the principles are relatively simple. Again, it starts with creating the largest, smoothest arc as you go through a corner. You start from the outside edge of the road and turn smoothly into the corner, apexing the inside edge of the road around the middle of the turn. You then allow the vehicle to track out all the way to the outside edge of the road accelerating as you exit the turn.

Fig.19

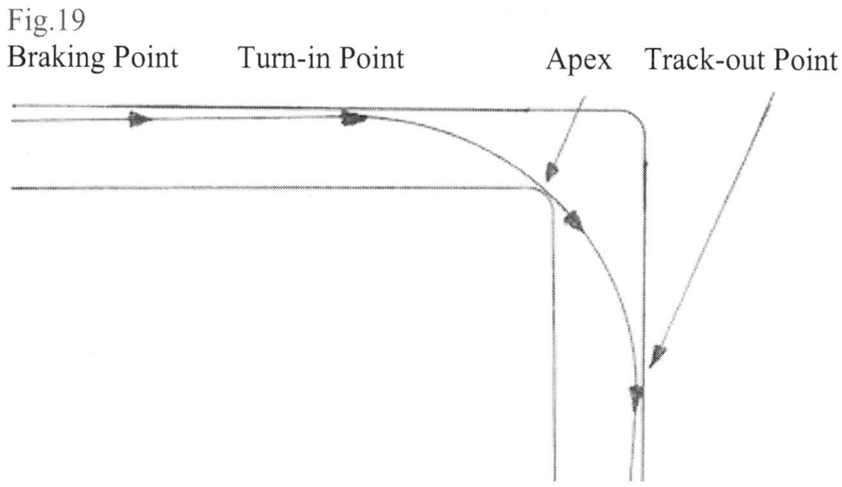

As I've mentioned before, if you were to *not* use the entire road available as indicated by the figure below you would have a tighter arc through the corner, producing slower speeds and putting more wear-n-tear on the tires.

Fig.20

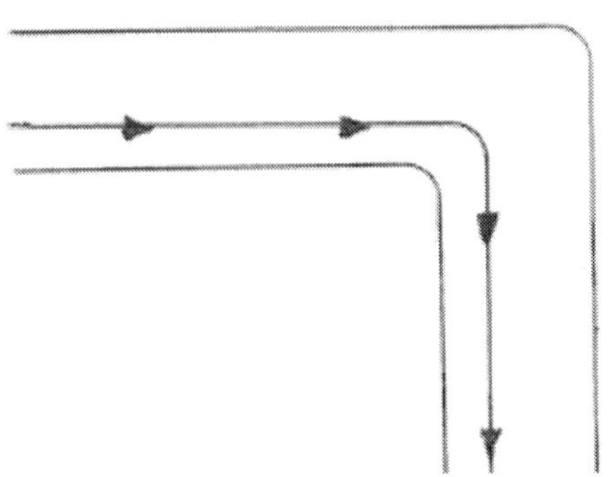

No two corners are alike. There are Increasing Radius Turns where the arc at the start of the turn is tight, but gradually opens up as you continue through the turn as in Fig. 21A. Conversely a Decreasing Radius Turn starts out with a wide arc but gradually tightens as you get further into the turn as in Fig. 21B.

Fig.21A Fig.21B

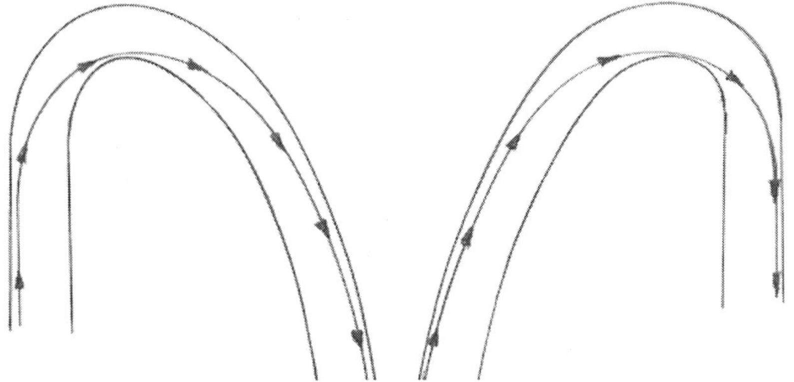

Another type of turn is the Compromised Corner. To understand this type you need to first understand the most important corner is the one that leads onto a straightaway. You want to try to accelerate at the earliest possible moment out of any corner; but this is particularly true when coming onto a straightaway where you can accelerate to your fullest. To do this, you once again must create the largest arc through the turn to maximize your exit speed. That means as you enter that turn you must be all the way on the outside edge of the track to maximize the arc. If there is a corner just before the corner that leads onto the straightaway you will have to compromise that first corner in order to be lined up for the second corner. You will have to use a tighter arc in the first corner with a very late apex to position yourself for the larger arc in the more important corner leading onto the straightaway.

Fig.22

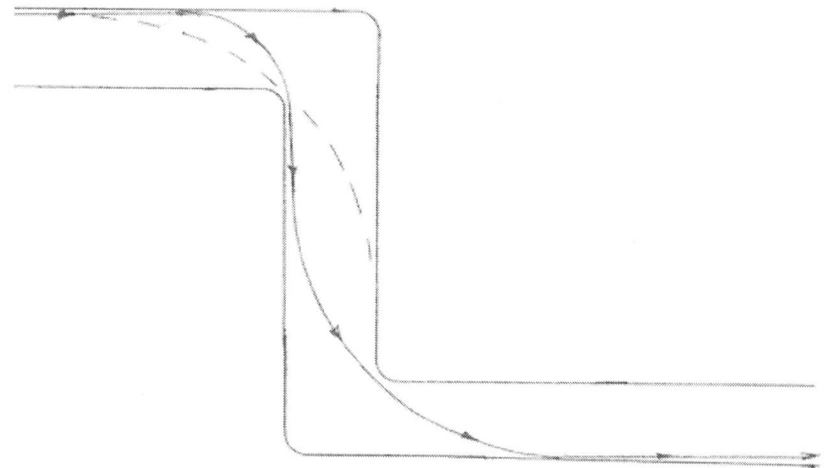

Put it all together, your fastest lap will feel like Sunday driving, feeling the rhythm of the road, just cruising around the track, the result of driving the Line and perfecting your race driving techniques.

The Secrets of a Race Car Driver:

*Look Ahead through the corner to tell your hands how much to turn the wheel.

*Look Ahead to better Feel if the car is Under or Over Steering.

*Look Ahead to Anticipate your Marks.

*Use Quick Hands to Correct and Recover any Under or Over Steer.

*Use your Peripheral Vision to keep tabs on your Competition.

*Have an Escape Route should a Competitor Lose Control in front of you.

*Slow In - Fast Out for the most efficient use of Momentum.

*Keep your Feet Following your Hands.

*Use Proper Feet, Hand and Seat Positions for Maximum Control.

*Threshold Brake for the shortest braking distances.

*Turn the wheel as little as possible for the largest, smoothest and fastest arcs.

*When in Doubt – Both Feet Out to maintain control of yourself and the car.

*Use the Knee to position your Foot on the pedals - avoiding Unintended Acceleration.

*Find the line – connect the dots, look ahead, use peripheral vision, see the entire track, all of the corner, feel the car, feel the condition of the track, be in tune with the car and the track – *be in the moment.* When you are fully given over to it, you enter the Zone – the place that athletes talk about – when you are completely connected to the moment, one with your racecar, one with the racetrack, not thinking, just *doing*.

This is the Zen of car racing.

Mark Donohue talked about it after winning the Indy 500: "...I don't even see it as a car. I just seem to be part of a well-oiled machine. That's the sensation that's so thrilling to me---knowing that everything in the system is working exactly as it's supposed to. ...There is no conscious skill involved at that point, and I feel what I'm doing is the obvious, natural thing to do. The driver is simply doing what the car wants to be done."#10

We lose sight of the *marvel* of traveling farther and faster than our feet can take us. Stuck in bumper-to-bumper traffic that's not hard to understand. But there is a joy to be found in driving a vehicle.

Even in heavy traffic, the Secrets of a Race Car Driver can help you better see the movement of cars around you, help you understand that being smooth with the controls keeps your car balanced on all four tires thereby giving you more control. Understanding how that balance is affected by acceleration, stopping and turning can add another dimension of satisfaction and awareness to your driving. Understanding the Line isn't and shouldn't be an attempt to get to the fastest speed but rather the *safest* speed, the one that blends you into the flow of traffic.

Using the Secrets of a Race Car Driver and discovering the Line in your daily driving can be seen, just like in racing, as very efficient driving.

It may not be as intense as racing and what Steve McQueen was referring to when he said, "It's life...everything else is waiting." But there is a satisfaction to be found in developing driving as a skill to be enjoyed, not just endured, feeling confident in maintaining control of your vehicle and your self no matter what the situation, a sense of accomplishment in feeling part of your vehicle, finding the rhythm of the road and feeling one with it, getting from the Green Flag to the Checkered Flag and home as smoothly and easily as possible.

TEACHING YOUR TEEN HOW TO DRIVE

Without having to hear: I know, I know, I know!

I've taught teen-driving programs at a number of venues and the number one complaint I get from the parents is always:

WHAT THE HECK ARE THEY THINKING??

Their teens are too close to the edge of the road or too close to the median strip or they accelerate too fast or too slow or they brake too late. Does any of this sound familiar? And if the parent tries to correct them the answer is always: "I know, I know, I know!" I've had parents freaked out because their teen drives angry. What's a parent to do??

Well, to help someone lost in the woods, you have to go to where they are. So, let's start by answering the question: what the heck are they thinking? And no, the answer is not, god only knows. The answer as it turns out is actually quite simple.

Remember all my talk about fight or flight, the reactions to fear?

Well, that is exactly what is happening.

Let's look at it from a teen's perspective. As children they were in the safety of family and friends and a familiar neighborhood. And once they got past the initial shock of their first day at school, school too had become a familiar and safe environment. However, when they become teens and learn to drive, they are now stepping out into an unknown and sometimes unsafe world. And it can be a daunting experience. It can, and more often than not *does*, elicit a fight or flight reaction.

In the case of the angry teen, it's fight. In the case of the teen that is too close to the edge of the road, it's flight. In both cases, they are looking right over the front of the car and no further. What is beyond that is unfamiliar and quite simply frightening.

Now, no teen is going to admit they are scared. That's where the "I know, I know, I know!" comes in. Relax parents don't take it personally. It's got nothing to do with you. I repeat: it's got nothing to do with you. It's just a fight reaction to fear. Call it a false sense of bravado, whistling in the dark. But don't *call* them on it or you will definitely get a much angrier reaction. And for those of you concerned about the anger? Stop and think about it. Your reaction to fear can go either way: fight or flight. And what better way to support the fight reaction than with anger? Again, it's just a reaction to fear.

So what's a parent to do?

When you buy new shoes you don't just take off running; you have to walk around in them first. Get the feel of them. Well, a car is no different.

My parents started me with backing the car up in the driveway about ten feet and then moving forward. And that was with a manual transmission and I was just big enough to reach the pedals. My first actual driving experience as a kid was my uncle let me drive around in a field with his old truck. We bounced all over the place and I had a blast! By the time I was old enough to get my license, I had already been driving up and down the street as well as backcountry roads.

Most parents don't have the luxury of wide-open fields and backcountry roads but the next best thing is a big and empty parking lot. And I do mean: BIG and EMPTY! Once again you want a parking lot with no lampposts or concrete parking space dividers. Big and Empty are the key words here. You want to put your teen in a place where they don't feel threatened. Allow them the opportunity to get the feel of the car without obstacles and distractions. Start off with a simple start and stop and then a back up and stop. Turn the wheel and go in a circle. Turn back in the other direction and go in a circle. Ask them if they can feel the car leaning as it turns. Move the wheel back and forth and feel how the car shifts its weight from one side to the other and how that effects how much the car wants to lean. Bring the speeds up gradually and see how the car feels. Just get used to the new shoes. Their confidence will grow as they get used to what their car *feels* like as they drive.

You want to work up to hard stops, and once they're comfortable they should learn what the A.B.S. brake feels like. When teens feel the pulsing of the A.B.S. brake they think they did something wrong and take their foot off the brake pedal. They have to learn all that grunting noise and stuttering sensation they are feeling is normal. It's the car doing its A.B.S. process and they are not hurting the car or doing anything wrong. Understanding this, they are better prepared should they ever *have* to use the A.B.S.

Each of these experiences needs to be spread out over time. And no expectations! Let the teen dictate the pace at which they learn but insist on keeping the speed down. And this is important, especially later when they are driving on their own, *you* the adult set the rules. Even to test the A.B.S. you don't have to be going any faster than thirty miles per hour. Set the rules from the beginning or it will just get more difficult later and *could* cost them their life!

It's amazing what we take for granted. Don't assume your teen knows what brake lights are indicating. Show them how to read the gas gauge and what the other gauges mean. Ask them what each switch on the dashboard does…besides the radio. Ask them if they know how to set the emergency brake? Show them the back up lights. If that fool in front of them at the traffic light accidently puts his car in reverse, know how to recognize back-up lights when they come on. They should definitely learn to leave plenty of room or get on the horn so that idiot doesn't plant himself in their grill when the light turns green!

When they first start out teens are going to be doing what comes naturally: they're going to be looking directly over the hood to what is just in front of the car. They are withdrawing. It's a flight reaction. The stuff we don't know about tends to intimidate us. So one way of dealing with it is to filter it out by withdrawing from it. Teens withdraw into their music plugged into their ears. They withdraw into their texting. They withdraw into their "Yeah, what ever." That withdrawal process is manifested behind the wheel by looking no further than the front of the car. They see only what's close to the car. As a result they hug the edge of the road or median strip or get too close to parked cars or brake too close to the cars that are stopped in front of them. A lot of times they will accelerate quickly to the speed limit because they are looking down at the speedometer. Everything about their driving is pulled back and in as if in a defensive position. Oh, and don't be fooled by the teen who speeds and gives you the old bravado-I-know-what-I'm-doing routine. That too is whistling in the dark – a milder fight reaction to fear.

So how do we fix it? We play a game called "eyes up." I talked earlier about "eyes up" at the race schools: always look ahead to where you *want* the car to go. So once your teen has gotten used to moving the car around that big empty parking lot, start pointing ahead to where you want them to go and get them to look where you're pointing. For instance, if they're going to make a U-turn they should be looking all the way out the side window. It tells their hands how much to turn the wheel since *where you look is where you steer*. It also helps them *feel* the car react, not *see* the car react.

Keep in mind your teen is experiencing everything for the first time. Dave Carman, who also teaches teen driving programs, talks about what I refer to as *road memory*. Similar to motor memory, it comes from repetition. With years of experience, *you* know how to judge braking distances, how much pressure to apply to the brake pedal, how fast to take that corner you go through everyday on your way to work. Your teen knows none of this. They don't have that bank of experience that allows them to have a "feel" for what they are doing. It is *all* new, and at times they may be clumsy. So, be patient.

Developing a "feel" for the vehicle by the way is the hardest thing for race driving students to learn. Until they can *feel* when the car is Under Steering or Over Steering they'll never drive as fast as the car is capable. And that won't happen until they learn to keep Looking Ahead. On the track I just keep pointing, showing them where they need to be looking; it eventually sinks in. Frustration aside, just keep pointing the way for *your* teen until they get comfortable with looking beyond the familiarity and safety of the hood.

Taking it out on the street is the next step. We've all seen the evening news filled with horrific car accidents. And every day we witness idiots behind the wheel vying for the Darwin Award. This is not lost on your teens. The public roads are a scary place. What your teens need to learn is that driving doesn't have to be a crapshoot. They do have an option that allows them to take control of their situation, an option that does *not* leave them at the mercy of the next Darwin Award winner. This is where the "eyes up game" comes into play.

The game is played out in two simple parts: As they pull out onto the street, point and *ask* them how far down the road they can see. It might be a case of several traffic lights. Ask them to describe what they see. Don't get discouraged when they say, there's a car in front of them, usually delivered with a certain amount of boredom - the fight reaction to fear. Ask them what else they see. If they complain they can't see around the vehicle in front of them, ask them if it's possible to see around the vehicle without getting out of their lane. You are getting them to figure it out for themselves. Keep them describing what's going on up ahead by asking them. Give them simple praise: that's good, very good, what else do you see? The idea is to keep them Looking Ahead at everything that is in front of them and to move their eyes around. Don't tell them to look in their mirrors, just ask them what is behind the car or along side of them? *That* will get them using their mirrors. This is key: always ask them, don't tell them. By asking, you are teaching them to be proactive. Again, they are figuring it out themselves. They are developing a process they can use anytime. Oh, and don't ask questions like, don't you see that car ahead? That's the same thing as telling them. The idea is to get them looking beyond the hood and engaging the world around them. Looking just over the hood they are reactive. Looking Ahead they are proactive. It doesn't matter they don't see that red light ahead you are looking at; just keep asking what they see ahead and praising them for it and eventually they will get to that red light that has you biting your lip.

Looking ahead also applies to going through corners. Get them to look all the way through the corner by asking them, how far can they see around the corner and what do they see? Use this opportunity to explain Where You Look is Where You Steer – your hands automatically follow your eyes. You can use the tree question from the Intro. When they Look Ahead all the way through the corner they are telling their hands how much to turn the wheel. Again, being proactive.

This, of course, is a gradual process. Start them out on the quietest road you can find, free from as much car and pedestrian traffic as possible, then gradually onto busier streets. The lesson is always the same: look as far ahead as possible, move your eyes around, check your mirrors.

The pedestrian they see at the curb or that car they see coming in from the side street: is he slowing down or is he texting and not paying attention? This is where the second part of the game comes in. It's called, "What if?"#11 What if this guy is texting and blows right through the light, what can you do? What if that pedestrian steps into the crosswalk? What if that dog takes off running out into the street? How can you protect yourself and protect them? What steps should you take just in case that guy coming into view *is* a bad driver? You are now teaching them to ANTICIPATE. By Anticipating what may be happening in front, behind and around them, they are creating a CUSHION OF SAFETY. And this Cushion allows them to take control of their situation. They are no longer at the mercy of bad drivers. It's a gradual process that, even for the most intimidated teen, teaches them by Looking Ahead and engaging what is going on around them they are taking control of what will happen to them as they continue down the road.

Once out on the highway, changing lanes presents an interesting dilemma. Even if you have your side mirrors set so you have a 360-degree view around the car, we all find ourselves looking over our shoulder just to make sure. This is where you have to remind your teen: Where You Look is Where You Steer. If you are going to look over your shoulder, you have to remember to keep the wheel straight. Concentrate on keeping your hands at 9 & 3. And don't look back over your shoulder until you are sure you have enough cushion in front of you in case the guy in front suddenly slows or stops.

Like I say, it's a gradual process; but raising a child you already know that. Oh, and don't be surprised if your teens play the game with *you*. What's good for the goose is good for the gander. Besides, that too will help them learn and you'll have a few laughs together along the way.

At the race schools when I had a student in one of our high performance vehicles, we had a kill switch we rested our hand on just in case. I generally gave them one shot at recovering the car if they got out of shape. After that, I just flipped the switch shutting off the engine. The car would then settle even though their foot was still on the gas pedal creating the problem in the first place. Once they regained control and got the car pointed in the right direction, I flipped the switch back on and we continued on down the track, still in one piece.

Recently, I've been surprised to find a growing number of teens have postponed getting their driver's license, particularly in cities where understandably there is better public transportation but also denser traffic that a new driver would find intimidating. For the "intimidated" teen and now that I think about it for the "overly enthusiastic" teen as well, the parent doing the training might want to consider knowing how to take control of the car from the right seat. If your teen's hands slip on the wheel, it might be necessary to grab the wheel to steady it. I reach over and adjust the steering of my race students if they are turning in too early or too late. But I warn them first by pointing where they should be looking. Don't grab the wheel unless it's an avoidance maneuver. They may fight you at the wrong time. Just remind them to Look Ahead where they Want the car to go. And point when necessary, that will adjust their steering.

If your teen's foot slips on the pedal or worse scenario they have a fight or flight reaction and jump on the gas instead of the brake, know how to take the car out of gear and eliminate unintended acceleration. I've had to do this with some of my racing students. With Automatic Transmissions you simply push the shifter from drive to neutral and can do so even if the car is accelerating. That will take the engine out of the equation and the car will settle and at least stop accelerating. And no it's not going to harm the transmission. Try it some time when *you* happen to be driving on an empty street. In the case of a Manual Transmission, it would be easier to just shut the engine off. With a key ignition, you would turn the key one click to shut off the engine while running, two clicks would lock the steering wheel and you don't want to do that! So get the feel of one click and do so again in a safe driving environment like a big empty parking lot. If it is a push button start, holding the push button in will shut off most engines even while accelerating. Practicing this on a safe street will tell you if your car is equipped with this safety feature. And again you want to make sure it doesn't lock the steering wheel.

Practicing all of these emergency maneuvers will give you a better feel for what to do should it be necessary and greater confidence going into the training session.

I can't talk about Teen Driving without talking about Distractions. More teens are killed while distracted than any other age group. The distractions are obvious: texting, cell phones, music, and just other teens carrying on. There are apps that a number of service providers offer that automatically shut off cell and texting capability and return a message that the person is currently driving. Most are triggered by the GPS and a specific speed usually around 25MPH.

You the adult are going to have to establish the parameters of what is permissible. Yeah, you want to be their buddy but you are also their parent. Establish parameters. A teen, or any kid for that matter, needs to know where the boundaries are. They will fight you but they *need* those boundaries.

Many states have laws governing the step-by-step training of teens until they are fully licensed. But once they are licensed you the adult need to continue to establish parameters such as how late they can be driving and when they are allowed to have other teens in the vehicle and how many.

The teen driving program sponsored by Toyota, required not only the teen taking the course but the parents as well, so the parents would know how to continue their teen's training. A contract between the teen and their parents was recommended that made teen *and parents* eliminate the use of cell phones for calls or texting while driving. And absolutely NO DRINKING and DRIVING, *that* includes not getting into a car with someone who has been drinking. Call home – NO QUESTIONS ASKED.

Something you may want to consider for you and your teen.

Finally, remember this:

THEY DO WHAT YOU DO, NOT WHAT YOU SAY.

MORE ABOUT DISTRACTIONS

It's getting to the point where we are living in a constant state of distraction. And it's taking the extreme experience of a thrill ride at Six Flags to get us into the current moment. In one of his few interviews, Woody Allen was asked what he believed in. He answered: "…the unending ability of people to be distracted."

It's easy to let your mind wander when you are on a long trip and on a wide-open stretch of highway. But just make sure, it's wide open. Make sure there's enough room around you and ahead of you on the highway that should something go wrong you have enough time to become aware and react safely. On crowded public streets or crowded highways and you're thinking about the honey-do list or the spat with the significant other or putting on makeup or fiddling with your IPod or cell phone or texting or…this list is endless…that's when you are a collision about to happen!

If you're not paying attention to your driving then you are a hazard; not only to your self but more importantly you are a hazard to others.

I know people like to think they can multitask and probably take great pride in it. But the truth of the matter is that when you multi-task while driving, it is the equivalent of drinking and driving! Regardless of what multitaskers think, studies have shown that multitasking leads to inefficiency.

The same thing is true with drinking and driving. Even one drink can slow down your reaction time and your cognitive abilities. So using the cell phone or texting while driving, regardless of how wonderful a multitasker you might think you are, you have just slowed down your ability to react to a situation on the road.

Look at it this way:

TEXTING IS WORSE THAN DUI.

At least the drunk is looking at the road.

This is not to make a joke out of or to downplay the seriousness of the CRIME of drunk driving, but realize that when you are texting you are NOT LOOKING AT THE ROAD AT ALL! And criminally so - like the drunk - you greatly increase the chances you will be killing that innocent family of four you are about to run into.

So you think you can talk on the cell phone even hands free and still drive? Answer this honestly: how many times have you been talking on the cell phone while driving and missed your turn?

I rest my case.

In racing there is a saying: “When the green flag drops, the bull#%#+ stops.” When you are racing you don’t have time to think about anything else; racing is the only thing that exists at that moment. You’re not thinking about mortgage payments, or soccer matches, or college tuitions. The rest of the world goes away. All you have time to think about is racing; and for good reason - your life depends on it. Every decision, every movement can very quickly become a life and death decision, a life and death movement. Like McQueen’s quote, “It’s life,” the experience is so powerful, so addictive, you very often feel more alive in those moments than at any other time in your life.

I’m not pointing this out to sell you on racing. We all have a romanticized view of racing. We all think of it as glamorous. What most people don’t realize is how much work goes into racing. Beside the physical work on the car I mentioned earlier, the driver has to prepare him or herself to study the track, know every turn, every change in elevation, every bump, every change in the surface of the track as the race goes on because temperature and the rubber from race tires adhering to the surface changes the track, not to mention the oil and other fluids that get spilled on the track. This is why they must practice over and over the techniques that allow them to maintain control of the racecar as everything around them is changing.

For safety sake, you might want to keep this level of concentration in mind when you get behind the wheel. The public street is not a racetrack and should never be used as such. But just like on a racetrack in an instant everything on the street can change. That car that swerves in front of you, the kid that runs in front of you, your car suddenly skidding - all of these will require your full concentration because your next decision may be life or death. By eliminating *your* distractions, you greatly increase your chance of making a life decision.

STREET RACING

For all you young guns that think that public-street racing is somehow more macho and lends an added element of danger and thrills, KNOW THIS: the danger is not to yourselves, the danger is to the innocent people around you.

The so-called "outlaw" racing you watch on TV or in the movies is "street" racing in name only. All those shows are shot under highly controlled circumstances and might just as well be racing on a designated racetrack. Believe me, the lawyers for those shows and movies make sure of that.

That warehouse district where you want to race at night? You think the danger is just yours. What about the guard who just got off his night shift and is trying to get home to his kids. You'd like to think it's his fault for pulling in front of you, but the truth is this: when he sees your headlights a block away, he assumes he has the five seconds it will take to cross the intersection. He can't see that you're traveling at over a hundred miles per hour and will be on him in two to three seconds, killing him instantly. The danger is not to you, the danger is to that sole provider for three kids. YOU are now responsible for whatever happens to his children!

All those friends you think you need to impress? If they are your friends, they'll go to a racetrack to cheer you on.

RACE CAR DRIVING IS NOT ABOUT IMPRESSING OTHERS!

IT IS ABOUT IMPRESSING YOUR SELF - by developing a skill set that expands your own knowledge and self-satisfaction.

If you are going to race, race on a racetrack. For racetracks near you Google on line or use SCCA.com as a good starting point. They are a national and regional race organization with a wide array of racing programs available.

SOME ADDITIONAL WORDS OF CAUTION

"RUNNING" WATER

We've all seen the destructive power of the tsunami that hit Japan. Moving water is immensely powerful. That little creek or river bed that overflowed and is now shooting across the roadway may not look like much but consider this: a gallon of water weighs a little over eight pounds. If that swollen creek is pushing a thousand gallons across the road, and that can often not look like much, you are looking at eight thousand pounds. And it's moving! That's the equivalent of an average truck. Your car weighs any where from three to five thousand pounds. In a confrontation, guess who is going to lose?

There are vehicles designed to ford rivers. They sit higher and the engine electronics are mounted in the upper half of the engine bay. But we're talking about shallow and passive water. Once you go above the axle line – half way up the wheels – you're asking for trouble. And if the river is "running;" forget it. Find some other route. Anyone who lives near mountains can tell you the danger of a river that is "running." Fed by snowmelt in the spring or a sudden rain, rivers and dry creek-beds suddenly become dangerous tracks for a fast moving train. Anything in its path will be swept away. Even the Los Angeles River, a benign cement trough that is usually a trickle, can become deadly. A sudden shower or during the rainy season and that trickle becomes a torrent. The evening news is then filled with the desperate attempts of brave rescue crews trying to snag some foolish kid who is being swept out to sea. If they miss them, they are never seen again.

DO NOT underestimate the power of fast moving water over a roadway. Find another route!

HYDROPLANING

The first thirty minutes of a rainfall are generally the most dangerous. All that oil, grease and dirt that have been gathering over the dry months with thousands of cars passing over the road surface are now going to be lifted up with the first rainfall. That stuff is no longer in the cracks, it's on the surface and we're talking slick like ice slick. Los Angeles is known for its countless sunny days. But when the first rain hits? The freeways become a pinball machine of bouncing cars ricocheting off each other. With the forecast of the first rain, there are sections of highway where tow trucks will line up, just waiting!

But don't panic. It doesn't mean pull over. JUST SLOW DOWN! By slowing down you are preventing a phenomenon known as HYDORPLANING.

Looking at the surface of your tires you see grooves cut into the surface of the rubber to form what are called treads. I know, some of you are sarcastically saying "really?" right now but hear me out. The grooves are designed to whisk away water, dirt and other debris from the surface of the tire so the "rubber can meet the road" to give you the necessary traction to control the car. Those grooves are designed to handle a certain amount of water. The faster you go the more water is being forced into those grooves. You can compress air but you can NOT compress water. So if you go fast enough to shove more water into the grooves than the grooves can disperse, then the water will lift the tire – and your three to five thousand pound car - off the road. This is called Hydroplaning. Again the fix is simple: SLOW DOWN!

The grooves forming the treads on a racecar's rain tire are huge by comparison. They have to whisk away a greater volume of water because of the higher speeds. And the rubber compound is much softer for better grip. But because of the softer rubber, the tires are designed to last only one race and in some cases only twenty or thirty laps. If the track starts to dry while on rain tires the tires will actually start to melt. The rubber will ball up forming what are called "marbles or clag" and the car becomes treacherous to handle forcing the driver to pit and change to new rains or a harder compound tire.

The tires on your car have to last thousands of miles, so the rubber is not as soft. Getting the right design and size of groove to whisk away water and dirt while keeping the right amount of rubber on the road for control is an exacting science that tire engineers are constantly trying to improve. Then too the rubber compound must also be correct: not too hard and not too soft. All season tires try to find that balance. There are specialty tires that are designed just for snow or off-road where you are dealing with lots of dirt and mud so the treads and grooves are bigger and the rubber compound softer. But those tires are best removed for regular-road use because they will wear out quicker.

Oh, and keep your tires properly inflated. If you've never done it before, ask your dealer, any mechanic or kid who likes cars and they can show you how. It takes a simple tire gauge that looks like a pencil with an eraser scale that slides up and down. The proper tire pressure is listed on the driver's door jam.

Take car of your tires, your tires will take care of you.

The same can be said of your entire car.

AIRBORNE AT WATKINS GLEN

Yeah, you knew sooner or later I was going to get to this.

I built my own racecar based on my friend Tim Abbott's design, a true craftsman who today is known for his prize-winning custom-built fly-fishing rods. Tim is shorter than me, so sticking my lanky frame into his design took a lot of re-engineering. It started with moving the gas tank from behind me in the cockpit to just under my out-stretched legs. It worked out fine because I was so reclined in the cockpit the gas tank doubled as the bottom of the seat that my legs rested against. It took a lot of re-triangulating of the chassis to get it all to fit under the body design. This car was an aerodynamic bullet with one of the lowest profiles in the class. It was, shall we say, a liberal reading of the rulebook. The first time I took it out on the track I could immediately feel the difference. The tachometer just kept climbing as I flew down the straightaway.

It took me a year to build it in my spare time, and most of the first season trying to sort it out and get everything on the chassis working optimally. Even at the last race of the season there was still some question about the rear spring rate. In high-speed corners there was some tail wagging, as the rear would slide out in Over Steer. It would freak out other drivers who were near me, but as Ron pointed out, that was a good thing. They wouldn't dare try to pass me in high-speed corners. Since on alternate weekends I was sliding around in Dave's dirt track racecar – he got hurt halfway through the season and I took over - getting a little sideways at over a hundred was not really an issue. Except of course for the wear and tear on the rear tires.

The last race was at the Glen, and I should have realized early on, the weekend was going downhill fast. My engine mysteriously shut off on the warm up lap of the qualifying session. I fixed the problem but missed the session and as a result was dead last in a field of thirty-eight cars. From back there, on a two-by-two rolling start I wouldn't be able to see the starter's flag station. This was before they installed the colored lights that tell you when the green flag has dropped and the race has started.

I watched the early races and saw the flagman was throwing an early green flag before the cars got out of shape from their side-by-side positions. He was throwing the start flag just as the lead cars came out of the right-hand turn onto the front straight.

As we rolled around the track two-by-two I was getting ready. As soon as the two lead cars went around the right-hander onto the front straight, I figured the green flag was waving and I nailed the gas pedal. I don't know what the others were waiting for; by the time I got to the right-hander I had cut the field in half. By Turn 1, I had so much momentum I slid into tenth place!

I knew that getting through the chicane in the Esses flat out would be critical if I were to have any chance of working my way up to the front-runners. Now that I look back on it, I guess I was a little crazy because I started blocking like crazy to slow down the field as we came out of Turn 1. I was trying to create a large-enough gap to the car in front of me, where he would be through the chicane by the time I got there. I would then have room to take the chicane flat out, pass him going up the hill and carry more speed onto the back straight.

A car started to edge past me. I squeezed on the gas, went up through the gears, and was headed for the Esses and the chicane at full throttle. I used my peripheral vision to get through the sweeping right-hander and had my eyes glued to the entry curb of the chicane. Just as I came out of the sweeper, the driver of the car in front must have panicked. He slammed on his brakes at the entry to the chicane! To this day I can still see the smoke coming off his tires. I looked left, then right and had nowhere to go! I was on him in a blink. My left front tire hit his right rear tire and I shot straight up in the air! I was looking at the sky. There was a moment of weightlessness. And then I was looking at the ground. All I could think to myself was, @%#t, I'm going to wreck my car!

The car hit, bounced and took out seven fence posts and a whole lot of chain link fence before finally coming to rest.

I climbed out from under the chain link fence that was draped over me and surveyed what remained of my car. I was broken hearted. All that work…

They came with a truck and a hook and gently lifted her by the roll bar and took her to my trailer. I packed up, went home, put her in the garage and closed the door. My racing days were over. As I've said, it doesn't take much money to go racing; it just takes all the money you've got. And at that point, I just didn't have enough to rebuild her.

Before the week was up, the word had gotten around. I got a call from Ron. He was kind enough to ask first if I would be racing the next season. Then, being the good Italian he is, he made me an offer I couldn't refuse. He offered to rebuild her…if he could have the car for the following season. I told him, "Come and get her."

He raced her the following year and had a great season. At Pocono Raceway, he took full advantage of the car's design during qualifying. He caught a draft on the back straight and another on the front straight putting him on the pole. He was a second and a half in front of the rest of the field. That's all well and good, but a second and a half normally covered the entire field, not first place to second. He had to lie about his qualifying time, saying it was slower than what the officials had recorded. He argued that the timers must have miscued because his mechanic had him at a slower time. Of course, that time was still fast enough to keep him on the "pole," first place for the start of the race.

This situation was a lot like passing on the highways in California. The speed limit was generally covered by what was called the "flow of traffic." You could pass, but just not too quickly or you would *draw attention* to yourself. We wanted the car to win; we just didn't want to draw attention from the officials.

He had a great season, but at the end of the year the governing body recognized the "liberal reading of the rules" and the "unfair" aerodynamic advantage the car had. They changed the rules and outlawed the car. I sold her a few years later to a hill-climb racer. My racing days were over.

Racing however is one of those addictions that never quite leave your system.

Now?

I teach.

A FINAL WORD

A car is just a tool. Like a hammer is an extension of your hand to bang in nails, a car is an extension of your legs used to cover greater distances in a shorter period of time. And though it is not the intended use, just as a hammer can be used as a weapon, a car too can be used just like a weapon.

On the racetrack you want to beat your fellow competitors, but you do it with skill and determination, *not* with unbridled anger. We've all heard of road rage. In its most extreme form drivers have been known to shoot other drivers in a fit of rage. In its lesser form though, road rage can be seen as a case of simply cutting someone off because we got angry or selfish that we didn't want anyone to pass us. How many times have you tried to pass someone and they sped up?

In racing you know what the other drivers want, they want to win. But out on the street, you don't know what the other driver is thinking or what they want. The guy who is trying to pass you may have a serious emergency: a wife about to give birth, a medical emergency of any kind whether they're a patient or the doctor. Or maybe, it's some young person being stupid. Or it's just some idiot. Get out of their way. You are not racing that person. The public street is *not* a racetrack, nor should you ever consider it that way. Leave life's anger, life's frustrations for the gym and a punching bag, *not* another person whether they are in a car or not.

Native American Indians in battle had a thing they called "Counting Coup." They would hit the other warrior on the back as they rode by, letting them know they could have just as easily killed them. It was a show of respect, a show of respect that bred respect. The warrior who had been hit in turn would respect that they owed their life to that warrior who let them live.

Haven't you been pleasantly surprised when someone pulled over and let you pass? And didn't you respect them for being considerate? After all, common courtesy can go a long way. Or were you driving with anger? Like the idiot that says, "That's right sucker! Get out of MY way!"

Remember this:

RESPECT IS A TWO-WAY STREET:
YOU HAVE TO GIVE RESPECT TO GET RESPECT.
EVERYTHING ELSE IS JUST FEAR.

Fear too can be seen as a two-way street. If you try to inflict fear in others through anger, it is because you yourself are afraid. Anger comes from fear. After all, what are the two main responses to fear? Flight or...*Fight.*

With that I leave you with a Race Car Driver's

FINAL SECRET

RESPECT YOUR FELLOW DRIVERS

ON THE ROAD.

They have as much right to be there as you.

And they too are just trying to get to the finish line.

NOTES

#1 ConceptcarZ.com/vehicle/z1133/tucker-48.aspx

#2 Ask.com – who invented ABS brakes.

#3 www.nhtsa.gov.

#4 answers.yahoo.com>AllCatagories>Cars&Transportation> Safety.

#5 HowStuffWorks "Airbag Inflation."

#6 www.carsdirect.com/car-safety-what-a-smart-airbag-is.

#7 Wikipedia – autonomous cruise control.

#8 Terry Earwood from one of his hilarious inspirational speeches.

#9 Toyota Driving Expectations, Parent Program Guide.

#10 Excerpted from *The Unfair Advantage* by Mark Donohue, © Bentley Publishers, Cambridge, MA, www.bentleypublishers.com, ISBN 0-8376-0069-3.

#11 Toyota Driving Expectations, Teen and Parent Driving Program.